MYCOTOXINS IN PLANT DISEASE

Mycotoxins in Plant Disease

Under the aegis of COST Action 835 'Agriculturally Important Toxigenic Fungi 1998–2003', EU project (QLK 1-CT-1998-01380), and ISPP '*Fusarium* Committee'

Edited by:

A. Logrieco, J.A. Bailey, L. Corazza and B.M. Cooke

Reprinted from *European Journal of Plant Pathology*, Volume 108, Issue 7, 2002

KLUWER ACADEMIC PUBLISHERS
DORDRECHT / BOSTON / LONDON

A C.I.P. Catalogue record for this book is available from the library of Congress

ISBN 0-4020-0871-6

Published by Kluwer Academic Publishers,
P.O. Box 17, 3300 AA Dordrecht, The Netherlands

Sold and distributed in the North, Central and South America
by Kluwer Academic Publishers,
101 Philip Drive, Norwell, MA 02061, U.S.A.

In all other countries, sold and distributed
by Kluwer Academic Publishers
P.O. Box 322, 3300 AH Dordrecht, The Netherlands

Printed on acid-free paper

Printed in the Netherlands

Contents

Foreword

There is an urgent need to expand our knowledge of both the nature of the toxigenic fungi that are widespread on economically-important plants and the effect of their toxic secondary metabolites on human health. Information about the production of mycotoxins by plant pathogens, particularly by species of *Fusarium*, *Aspergillus* and *Penicillium*, their occurrence in infected plants, as well as their role in the plant-pathogen interaction, for example as virulence/pathogenicity factors, is a pre-requisite for preventing plant disease and hence for reducing the levels of mycotoxin contamination. *Fusarium* infections in cereals and other crops are a particular problem world-wide and recent epidemics on wheat in Europe, the USA and Canada have again focused attention on this problem. Furthermore, species of *Aspergillus* and *Penicillium* and their related mycotoxins, particularly Ochratoxin A, represent another consistent problem on cereals and grapes, especially in Europe where 40% of the global grape crop is grown.

The aim of this publication is to gather together specialist updated reviews based on papers originally presented during a Workshop of EU Cost Action 835 entitled 'Agriculturally Important Toxigenic Fungi', held in Rome, 7–8 October 1999 at the Plant Pathology Research Institute. We hope the diversity of the contents will stimulate discussion, encourage the sharing of information and result in cross-fertilization of ideas needed for the solution of the present problems. This special issue will be of particular value to interdisciplinary scientists and especially mycologists, mycotoxicologists, plant pathologists and those concerned about the quality of food and food products.

We express our sincere appreciation to the contributors for their excellent and timely contributions. We are also extremely grateful to Dr John Bailey and Professor Mike Cooke for their excellent and fruitful reviews of all the papers, to Mariella Quarto for careful support in preparing the original text and to Dr Roel Mulder for supporting the project.

ANTONIO F. LOGRIECO
LUCIANA CORAZZA
ANTONIO BOTTALICO

European Journal of Plant Pathology **108:** 597–609, 2002.

Toxigenic *Fusarium* species and mycotoxins associated with maize ear rot in Europe

A. Logrieco, G. Mulè, A. Moretti and A. Bottalico
Institute of Sciences of Food Production, CNR, Viale L. Einaudi 51, 70125, Bari, Italy (Fax: +390805486063; E-mail: a.logrieco@area.ba.cnr.it)

Key words: maize pink ear rot, *Fusarium* mycotoxins, *Fusarium graminearum*, zearalenone, trichothecenes, deoxynivalenol, *Fusarium verticillioides*, fumonisins, *Fusarium proliferatum*, fusaproliferin, moniliformin

Abstract

Several *Fusarium* species occurring worldwide on maize as causal agents of ear rot, are capable of producing mycotoxins in infected kernels, some of which have a notable impact on human and animal health. The main groups of *Fusarium* toxins commonly found are: trichothecenes, zearalenones, fumonisins, and moniliformin. In addition, beauvericin and fusaproliferin have been found in *Fusarium*-infected maize ears. Zearalenone and deoxynivalenol are commonly found in maize red ear rot, which is essentially caused by species of the *Discolour* section, particularly *F. graminearum*. Moreover, nivalenol and fusarenone-X were often found associated with the occasional occurrence of *F. cerealis*, and diacetoxyscirpenol and T-2 toxin with the occurrence of *F. poae* and *F. sporotrichioides*, respectively. In addition, the occurrence of *F. avenaceum* and *F. subglutinans* usually led to the accumulation of moniliformin. In maize pink ear rot, which is mainly caused by *F. verticillioides*, there is increasing evidence of the wide occurrence of fumonisin B_1. This carcinogenic toxin is usually found in association with moniliformin, beauvericin, and fusaproliferin, both in central Europe due to the co-occurrence of *F. subglutinans*, and in southern Europe where the spread of *F. verticillioides* is reinforced by the widespread presence of *F. proliferatum* capable of producing fumonisin B_1, moniliformin, beauvericin, and fusaproliferin.

Abbreviations: AcDON – Mono-acetyldeoxynivalenols (3-AcDON, 15-AcDON); AcNIV – Mono-acetylnivalenol (15-AcNIV); BEA – Beauvericin; DiAcDON – Di-acetyldeoxynivalenol (3,15-AcDON); DAcNIV – Diacetylnivalenol (4,15-AcNIV); DAS – Diacetoxyscirpenol; DON – Deoxynivalenol (Vomitoxin); FB_1 – Fumonisin B_1; FB_2 – Fumonisin B_2; FB_3 – Fumonisin B_3; FUP – Fusaproliferin; FUS – Fusarenone-X (4-Acetyl-NIV); FUC – Fusarochromanone; HT2 – HT-2 toxin; MAS – Monoacetoxyscirpenol; MON – Moniliformin; NEO – Neosolaniol; NIV – Nivalenol; T2 – T-2 toxin; ZEN – Zearalenone; ZOH – zearalenols (α and β isomers).

Introduction

Several *Fusarium* species are widespread pathogens on maize in temperate and semi-tropical areas, including all European maize-growing areas. They cause root, stem and ear rot, with severe reductions in crop yield, often estimated at between 10% and 30%. In addition, certain strains are capable of producing mycotoxins which can be formed in pre-harvest infected plants, or in stored grains. The phase of maize fusariosis with the greatest mycotoxicological concern is the ear rot, but the formation of mycotoxins in rotted stalks (notably ZEN, ZOH and DON) (Bottalico et al., 1985; Lew et al., 1997), infected leaves (NIV) (Lew et al., 1997), and in whole plants (ZEN) (Oldenburg, 1993) could represent a significant risk for forage and silo maize. The occurrence of mycotoxins in maize kernels is of great concern worldwide, because their presence in feeds and foods is often associated with chronic or acute mycotoxicoses in livestock and, to a lesser extent, in humans. It has been estimated that 25% of the world food crops is affected by mycotoxins (Charmley et al., 1995), but for some *Fusarium* toxins in maize, such as DON and FB_1, it is likely that this percentage is even higher (Eriksen and Alexander, 1998; Bullerman, 1996).

In this review, the most relevant aspects of the distribution of toxigenic *Fusarium* species and related mycotoxins in ear rot of maize in European countries are reviewed, with emphasis on the occurrence of trichothecenes, zearalenone, and fumonisins in infected plants standing in the field. Moreover, the relationship between the ecological distribution of *Fusarium* species and the relative mycotoxin profile is also stressed, because this may be useful for predicting the mycotoxins that are most likely to be formed in maize and those which may be affected by conducive factors such as tillage practices, host genotypes and environmental conditions.

Fusarium species involved and mycotoxins produced

The species of *Fusarium* causing maize ear rot are worldwide in distribution and are characterized by the co-occurrence or by the quick succession of different species. It is common to isolate up to nine different *Fusarium* species from a single fragment of infected tissue or kernel. Together with the restricted number of toxigenic species which are regarded as pathogenic, there are also several less pathogenic or opportunistic *Fusarium* species capable of producing considerable amounts of toxins. The toxigenic profile of a contaminated crop is therefore determined not only by the predominant pathogenic species, but also by the opportunistic species (Bottalico, 1997).

Fusarium species are responsible for at least two kinds of maize ear rot, roughly differentiated as red ear rot or red fusariosis, mainly caused by species of the *Discolour* section, and pink ear rot or pink fusariosis mainly caused by representatives of the *Liseola* section (Shurtleff, 1980; Chelkowski, 1989). The predominant species causing maize red ear rot are *F. graminearum* (teleomorph *Gibberella zeae*), *F. culmorum*, *F. cerealis* (syn. *F. crookwellense*), and *F. avenaceum* (teleomorph *G. avenacea*). The species frequently isolated from maize pink ear rot are essentially the widespread anamorphs of the rather rare *Gibberella fujikuroi*, namely, *F. verticillioides* (syn. *F. moniliforme*), *F. proliferatum*, and *F. subglutinans*. Among the other toxigenic *Fusarium* species less frequently isolated from both types of maize ear rot are *F. equiseti* (teleomorph *G. intricans*), *F. poae*, *F. sporotrichioides*, *F. acuminatum* (teleomorph *G. acuminata*), *F. semitectum* (syn. *F. pallidoroseum*, *F. incarnatum*), *F. solani* (teleomorph *Nectria haematococca*), and *F. oxysporum*. Finally, there are other species which are only sporadically isolated from maize, such as *F. anthophylum*, *F. chlamydosporum* (syn. *F. fusarioides*), *F. compactum*, *F. heterosporum* (syn. *F. reticulatum*, *F. graminum*), *F. lateritium*, *F. sambucinum*, *F. torulosum*, and *F. venenatum*. It is worth stressing that the complex species *G. fujikuroi* has been subdivided into seven distinct mating populations (biological species), indicated from A to G, and covering ten *Fusarium* anamorphs (Leslie, 1995). Those most frequently found on maize were *F. verticillioides* (A), *F. proliferatum* (D), and *F. subglutinans* (E), which can also be differentiated by their toxigenic capabilities (Moretti et al., 1997).

Mycotoxins produced

Fusarium species on maize can produce many mycotoxins, some of which are of notable importance. The naturally occurring *Fusarium* mycotoxins belong to the trichothecenes, zearalenones, and fumonisins. Moreover, MON, BEA and FUP have also been found in naturally infected maize kernels and are considered as emerging toxicological problems. The mycotoxins produced by *Fusarium* species from cereals are illustrated in Table 1.

Trichothecenes

Of several trichothecene derivatives produced by *Fusarium* species, only a few have been encountered as natural contaminants of maize products. The *Fusarium* trichothecenes have been divided into type A-trichothecenes, characterized by a functional group other than a ketone at C-8, and type B-trichothecenes with only the carbonyl at C-8. The type A-trichothecenes include: (a) T2 and HT2, mainly produced by strains of *F. sporotrichioides*, *F. acuminatum*, and *F. poae*; (b) DAS, and MAS, chiefly produced by strains of *F. poae*, *F. equiseti*,

Table 1. Mycotoxigenic *Fusarium* species associated with cereals and their mycotoxins

Fusarium species[a]	Mycotoxins[b]
F. acuminatum	**T2**, **MON**, HT2, DAS, MAS, NEO, BEA
F. anthophilum	BEA
F. avenaceum	**MON, BEA**
F. cerealis	**NIV**, **FUS**, **ZEN**, ZOH
F. chlamydosporum	**MON**
F. culmorum	**DON**, **ZEN**, **NIV**, FUS, ZOH, AcDON
F. equiseti	**ZEN**, **ZOH**, MAS, DAS, NIV, DAcNIV, FUS, FUC, BEA
F. graminearum	**DON**, **ZEN**, **NIV**, **FUS**, **AcDON**, DAcDON, DAcNIV
F. heterosporum	**ZEN**, ZOH
F. nygamai	BEA, FB_1, FB_2
F. oxysporum	**MON**, BEA
F. poae	**DAS**, **NIV**, **FUS**, MAS, T2, HT2, NEO, BEA
F. proliferatum	**FB_1**, **BEA**, **MON**, **FUP**, FB_2,
F. sambucinum	**DAS**, T2, NEO, ZEN, MAS, BEA
F. semitectum	**ZEN**, BEA
F. sporotrichioides	**T2**, **HT2**, **NEO**, MAS, DAS
F. subglutinans	**BEA**, **MON**, FUP
F. tricinctum	**MON**, BEA
F. verticillioides	**FB_1**, FB_2, FB_3

[a]*Fusarium* nomenclature according to Nelson et al. (1983). [b]Bold letters indicate the important mycotoxins. *Abbreviations*: AcDON – Mono-acetyldeoxynivalenols (3-AcDON, 15-AcDON); AcNIV – Mono-acetylnivalenol (15-AcNIV); BEA – Beauvericin; DiAcDON – Di-acetyldeoxynivalenol (3,15-AcDON); DAcNIV – Diacetylnivalenol (4,15-AcNIV); DAS – Diacetoxyscirpenol; DON – Deoxynivalenol (Vomitoxin); FB_1 – Fumonisin B_1; FB_2 – Fumonisin B_2; FB_3 – Fumonisin B_3; FUP – Fusaproliferin; FUS – Fusarenone-X (4-Acetyl-NIV); FUC – Fusarochromanone; HT2 – HT-2 toxin; MAS – Monoacetoxyscirpenol; MON – Moniliformin; NEO – Neosolaniol; NIV – Nivalenol; T2 – T-2 toxin; ZEN – Zearalenone; ZOH – zearalenols (α and β isomers).

F. sambucinum, and *F. sporotrichioides*; and (c) NEO, mainly produced by strains of *F. sporotrichioides*, *F. poae*, and *F. acuminatum*. The type B-trichothecenes include: (a) DON, and its derivatives mono- (3- and 15-AcDON), and diacetylated (3,15-AcDON) derivatives, produced by strains of *F. graminearum* and *F. culmorum*; and (b) NIV and FUS, and diacetylated derivatives (4,15-AcNIV), produced by strains of *F. cerealis*, *F. poae*, *F. graminearum* and *F. culmorum*. Trichothecenes cause a variety of toxic effects in laboratory and farm animals, including skin inflammation, digestive disorders, haemorrhages in several internal organs, haemolytic disorders and depletion of the bone marrow, impairment of both humoral and cellular immune responses, and nervous disorders (Rizzo et al., 1992; IARC, 1993; Rotter and Prelusky, 1996). Trichothecenes are commonly associated with several mycotoxicoses in livestock, including haemorrhagic and emetic syndromes; and have also been implicated in human toxicoses (Bhat et al., 1989; IARC, 1993; Wang et al., 1993; Beardall and Miller, 1994; Ehling et al., 1997).

Zearalenones

Zearalenone (ZEN) is mainly produced by *F. graminearum*, *F. culmorum*, *F. cerealis*, *F. equiseti*, and *F. semitectum*, often associated with ZOH (α- and β-zearalenol isomers). ZEN is among the most widely distributed *Fusarium* mycotoxin in agricultural commodities, and has very often been encountered, even at very high concentrations, in maize. ZEN is an uterotrophic and estrogenic compound responsible for recurring toxicoses in livestock, characterized by hyper-estrogenism in swine, and infertility and poor performance in cattle and poultry (Kuiper-Goodman et al., 1987; IARC, 1993).

Fumonisins

Fumonisins were first isolated from *F. verticillioides*, and then found in cultures of *F. proliferatum* and in a few other less ecologically important *Fusarium* species (Nelson et al., 1992). Four series of fumonisins have been described, and named A, B, C, and P. The

B series includes the most active fumonisins, particularly FB_1 which is causing great mycotoxicological concern (Hopmans and Murphy, 1995; Musser et al., 1996; Abbas et al., 1998). Fumonisin B_1 occurs naturally at biologically significant levels in maize and in a variety of maize-based human foodstuffs and animal feeds worldwide (Rheeder et al., 1992; Sanchis et al., 1995; Logrieco et al., 1995; Shephard et al., 1996). Feeds contaminated with FB_1 are known to cause leukoencephalomalacia in horses (ELEM), pulmonary oedema and hepatic syndrome in swine (PPE), poor performance in poultry, and alteration in hepatic and immune function in cattle (Ross et al., 1990; Kellerman et al., 1990; Harrison et al., 1990; Osweiler, 1995). The first case of ELEM was reported in north Italy associated with the presence of FB_1 in maize-based feed (Caramelli et al., 1993). Moreover, home-grown corn contaminated with FB_1 has been statistically associated with high rates of human oesophageal cancer (HEC) in areas of southern Africa, China, and the southeastern U.S.A. (Marasas, 1995). A similar claim has been made for northeastern Italian areas (Franceschi et al., 1990), where FB_1 was found in maize flour (polenta) in concentrations up to 3.76 mg kg^{-1} (Pascale et al., 1995). The evidence that cultures of *F. verticillioides* and samples of FB_1 are capable of promoting liver cancer in rats (Gelderblom et al., 1996), together with other observations obtained from several genotoxic tests, led IARC (1993) to classify the toxins of *F. verticillioides* as carcinogenic to animals and possibly as carcinogenic to humans (Group 2B).

Moniliformin

Moniliformin has been purified from cultures of several *Fusarium* species, including: *F. subglutinans*, *F. proliferatum*, *F. avenaceum*, *F. tricinctum* and several others of less ecological importance (Marasas et al., 1986; Chelkowski et al., 1990; Schütt et al., 1998). Diets containing culture material naturally contaminated with MON, or amended with purified MON, were responsible for reduced performance, haematological disorders, myocardial hypertrophy, and mortality in rodents, chicks, ducklings, and pigs (Ledoux et al., 1995; Harvey et al., 1997).

Beauvericin

Beauvericin is a well known cyclic hexadepsipeptide, first reported to be produced by some entomopathogenic fungi (*Beauveria* spp.) and then found in cultures of strains of *F. semitectum*, *F. subglutinans* and *F. proliferatum*, isolated from maize and maize-based feed for swine (Plattner and Nelson, 1994; Moretti et al., 1994; 1995). In addition, BEA was found in cultures of several other *Fusarium* species, including *F. avenaceum*, *F. acuminatum*, *F. equiseti*, *F. poae* and *F. oxysporum* (Logrieco, 2000). BEA was detected for the first time in maize ear rot in Poland (Logrieco et al., 1993), and then found as a toxic contaminant of maize in Italy (Bottalico et al., 1995; Logrieco et al., 1995). Besides its high toxicity to insects (Gupta et al., 1991), BEA is also cytotoxic to mammalian cell tissues, and was reported to cause apoptosis in both murine and human cell lines (Macchia et al., 1995). Moreover, BEA showed toxic effects on the contractility of guinea pig smooth muscle (Krska et al., 1997).

Fusaproliferin

Fusaproliferin is a novel sesterterpene first purified from a culture of *F. proliferatum* from maize ear rot in northern Italy (Ritieni et al., 1995), and then found both in cultures of several strains of *F. proliferatum* and *F. subglutinans* (Moretti et al., 1997), and in naturally infected maize (Ritieni et al., 1997a). Investigations on the toxicity of FUP indicated that this toxin is lethal to larvae of *Artemia salina*, and cytotoxic to the SF-9 insect cell line and to IARC/CL 171 human B lymphocyte culture (Logrieco et al., 1996). Moreover, high mortality in broiler chicks fed with maize cultures of *F. proliferatum* was reported by Javed et al. (1993) and Ramakrishnan and Wu (1994), and severe teratogenic effects were observed in chick embryo bioassays by Ritieni et al. (1997b).

Ear rot of maize in Europe

Fusarium *species in ear rot of maize*

The relative incidence of *Fusarium* species and related mycotoxins found in maize ear rot diseases in Europe are tentatively grouped into red fusariosis and pink fusariosis, and summarized in Table 2. The distribution and the prevalence of different *Fusarium* species causing the two kinds of ear rot disease are largely governed by environmental conditions, primarily temperature, as well as by many other factors including agro-technical practices (Arino and Bullermann, 1994). In general, red

Table 2. Toxigenic *Fusarium* species and associated mycotoxins from maize ear rot in Europe

Species[a]	Incidence		Mycotoxin found[b]
	North/Centre	South	
Red ear rot or Red fusariosis			
F. graminearum	+++	+	**DON**, AcDON, **NIV**, FUS, **ZEN**
F. subglutinans	++	±	**MON**, BEA, FUP
F. avenaceum	++	±	**MON**
F. cerealis	+	±	**NIV**, FUS, **ZEN**, ZOH
F. culmorum	+	−	**DON**, NIV, **ZEN**, ZOH
F. sporotrichioides	+	−	**T2**, **HT2**, NOS
F. poae	+	−	**DAS**, **NIV**
F. equiseti	+	±	DAS, ZEN, ZOH
F. acuminatum	+	±	T2, NEO
F. verticillioides	+	+	—
F. proliferatum	+	+	—
Pink ear rot or Pink fusariosis			
F. verticillioides	+	+++	$\mathbf{FB_1}$, $\mathbf{FB_2}$, FB_3
F. proliferatum	±	+++	$\mathbf{FB_1}$, FB_2, **FUP**, **MON**, **BEA**
F. subglutinans	+++	+	**MON**, **BEA**, FUP
F. graminearum	+	±	—
F. culmorum	+	±	—
F. equiseti	+	±	—
F. solani	±	+	—
F. semitectum	±	+	—
F. cerealis	±	±	—
F. sporotrichioides	±	−	—
F. oxysporum	−	+	—

For footnotes refer Table 1.

fusariosis is particularly severe in years and locations characterized by frequent rainfall and low temperatures during the summer and early fall (Ellend et al., 1997; Bočarov-Stančić et al., 1997), while pink fusariosis prevails in drier and warmer climates of southern areas (Bottalico et al., 1986; 1995; Bottalico and Logrieco, 1988). In the last ten years, the most frequently reported dominant species was *F. verticillioides* associated with *F. subglutinans*, and followed only to a slightly lesser extent by *F. graminearum*. Furthermore, the emerging trend of *F. proliferatum* moving from southern to northern maize-growing areas is well documented (Bottalico, 1997; 1998).

Maize red ear rot

In maize red ear rot, the pathogen, notably *F. graminearum*, usually starts from the tip of the ear and develops a reddish mould covering the ear extensively. The brownish perithecia of the teleomorph *G. zeae* are commonly observed on the infected husks and ear shanks. A comprehensive evaluation of reports from Austria (Krska et al., 1996; 1997; Lew et al., 1991), Slovenia (Milevoj, 1997), Yugoslavia (Bočarov-Stančić et al., 1997; Lević et al., 1997), Poland (Chelkowski, 1989; Lew et al., 1996); Czech Republic (Nedelnik, 2000), and România (Nagy and Bâgiu, 2000), clearly indicates that *F. graminearum* is increasingly distributed from central to northern European areas, and is occasionally associated with many additional *Fusarium* species, the occurrence and prevalence of which changes from region to region, and year to year, depending mainly on the climatic conditions (temperature and rain) and tillage practices (crop rotation, fertilization and planting area). In this context a very relevant role is played by host genotypes (maturity class, *Fusarium*-susceptibility) (Doko et al., 1985; Visconti, 1996). There are some other *Fusarium* species associated with *F. graminearum*, namely *F. subglutinans*, which predominates over *F. graminearum* in some Austrian locations (Lew et al., 1991), *F. culmorum* and *F. cerealis*, which were found to be more common in central European areas, and *F. avenaceum*, the relative presence of which seems to increase from central to northern European areas. Besides the more representative species, including *F. graminearum*, *F. subglutinans*, *F. cerealis*, *F. culmorum* and *F. avenaceum*, which represent almost

90–95% of all species isolated, *F. sporotrichioides*, *F. poae*, *F. equiseti*, and *F. acuminatum*, and to a lesser extent *F. verticillioides* and *F. proliferatum*, were also isolated.

The *Fusarium* species profile, and consequently mycotoxin accumulation can change drastically with insect damage. Lew et al. (1991) noted a striking influence of European corn borer injuries on the *Fusarium* species profile on maize in Austria in 1988–89. On maize ears not damaged by the corn-borer, *F. graminearum* and *F. subglutinans* were evenly represented (54.5% and 48.7%, respectively), while on ears damaged by the corn-borers there were many more species belonging to the *Liseola* section (*F. subglutinans*, *F. verticillioides*) (up to 80%) than to the *Discolor* section (*F. graminearum*, *F. culmorum*, *F. cerealis*) (less than 15%). Thus, control of the European corn-borer can lead to a reduction of MON and FB_1 produced by *Liseola* representatives, but does not affect the occurrence of ZEN, DON and NIV which are produced by *Fusarium* species belonging to the *Discolor* section.

Maize pink ear rot

In maize pink ear rot, the pathogens colonize the ear from the tip by external infections, but the kernels may also be infected through the silk at the silk stage to make ear colonization complex. In fact, the most important pathway for *F. verticillioides* to achieve seed-borne kernel infection is through silk infections at the silk stage (Munkwold et al., 1997). Thus, in association with a more generalized colonization originating from airborne inoculum colonizing the ears from the tip, a random kernel rot phase of the disease can also occur, appearing as randomly scattered individuals or groups of kernels, usually tan to brown, which develop pink mycelium under wet conditions. In this case, the risk of mycotoxin (fumonisin) accumulation in infected kernels may begin early in maize ear development and increases as the kernels reach physiological maturity (Warfield and Gilchrist, 1999). Maize pink ear rot is commonly observed from southern to central European areas, and the species most frequently isolated is *F. verticillioides*, associated with *F. subglutinans* and to a lesser extent with *F. proliferatum*. *Fusarium proliferatum* is more common in southern European areas, but it is displaced by *F. subglutinans* in central areas where the latter predominates as the maize ear rot agent and is usually isolated in much higher amounts than *F. verticillioides*. *Fusarium proliferatum* was commonly reported together with *F. verticillioides* in Italy (Logrieco et al., 1995), but in Austria (Krüger, 1989; Krska et al., 1997) Croatia (Jurjević et al., 1997), Slovak Republic (Piecková and Jesenská, 1997; Srobárová, 1997), Hungary (Szécsi, 1994), and especially in Poland (Kostechi et al., 1995) the occurrence of *F. proliferatum* was rarely recorded. However, it seems that unusually drier and warmer summers, like those prevailing in the 1990s led to an increase of *F. proliferatum* in central Europe, as reported both for the Slovak Republic in 1996 (Srobárová et al., 2000), and in Austria where the overall number of *F. proliferatum* infections rose from less than 1% in the 1980s to 2–11% towards the end of the 1990s, leading to an expected increase of fumonisin in contaminated samples (Adler et al., 2001). Associated with *F. verticillioides*, *F. subglutinans* and *F. proliferatum*, many other species were commonly isolated from maize pink ear rot. The species incidence reported in Yugoslavia by Lević et al. (1997) during a three-year survey (1994–96) of freshly harvested maize ears, included *F. verticillioides* (63%), *F. subglutinans* (51%), *F. graminearum* (12%), *F. proliferatum* (10%), *F. oxysporum* (6%), and *F. solani* (2%). Among the other less frequently isolated species were *F. equiseti*, *F. sporotrichioides*, *F. chlamydosporum*, *F. cerealis*, and *F. semitectum*. In Italy, where environmental conditions are often conducive to a high incidence of maize pink ear rot, *F. verticillioides* predominates (in almost 100% of the infected kernels), with a profuse co-occurrence of *F. proliferatum* in southern areas (in over 60% of the infected kernels), then decreasing from central to northern areas (in about 54% and 34% of kernels) (Logrieco et al., 1995; Ritieni et al., 1997a). The higher occurrence of *F. proliferatum* in Italy has focussed attention on this species, which is usually confused with other closely related species of the *Liseola* section, and led to a more correct evaluation of its pathogenic and toxigenic capabilities (Logrieco et al., 1995). A spreading trend of *F. proliferatum* similar to that observed in Italy, was reported for nearby European countries (Lević et al., 1997; Srobárová, 1997).

Occurrence of zearalenone and deoxynivalenol

Epidemics of maize red ear rot, induced by *F. graminearum* and *F. culmorum* usually lead to the occurrence of ZEN, DON, and 3AcDON, as reported in Poland (Chelkowski, 1989; Grabarkiewicz-Szczesna et al., 1996), and Yugoslavia (Bočarov-Stančić et al., 1997). In relation to the severity of the disease, it is possible to find unexpectedly high concentrations of toxins. In maize ear samples collected in 1988–89 in Austria, mostly infected by *F. graminearum*, *F. culmorum* and *F. cerealis*, Lew et al. (1991) found very high amounts of ZEN ($40\,mg\,kg^{-1}$), DON ($500\,mg\,kg^{-1}$) and NIV ($10\,mg\,kg^{-1}$). More recently, a survey of 85 freshly harvested maize ear samples, collected in 1996 in the eastern part of Austria, mostly contaminated by *F. graminearum*, contained DON in 95% of samples, 15-AcDON in 54%, 3-AcDON in 3.5% and ZEN in 70% (Ellend et al., 1997). A similar situation, but with much greater amounts of DON, 15-AcDON, 3-AcDON, and ZEN, was found for samples of maize ears infected by *F. graminearum*, collected from 1988 to 1991 in Poland (Visconti et al., 1990; Grabarkiewicz-Szczesna et al., 1996). Additional observations carried out in Poland on the distribution of mycotoxins in maize ears infected by *F. graminearum*, revealed that the levels of DON, ZEN, 3-AcDON, and 15-AcDON in whole ears were not evenly distributed, but were greater in the axial stems than in the kernels (Perkowski et al., 1991). In a survey carried out in 1977 in southern Italy (Metapontum), ZEN was found in *F. graminearum*-infected ears of maize sown as a second crop and harvested at the end of November (Bottalico, 1979), while high concentrations of DON were found in almost all the selected *F. graminearum*-infected maize ear samples (95%) collected in 1987 in northern Italy (Lombardy) (Bottalico et al., 1989).

Occurrence of nivalenol and fusarenone

Nivalenol and fusarenone have often been reported in maize red ear rot all over the European maize growing areas. Their formation in infected ears may be due to NIV-chemotypes of *F. graminearum*, especially when found together with DON and ZEN, and in the absence or scant presence of *F. cerealis*, as reported for Romania (Moldavia) (Ciudin and Bazgan, 1991), Italy (Logrieco et al., 1992), and Hungary (Szécsi and Bartok, 1995). But, the occurrence of NIV and FUS in European areas appeared to be related more to the spread of *F. cerealis* than that of *F. graminearum*. In fact, strains of *F. cerealis* from red ear rot of maize were essentially able to produce NIV and FUS associated with ZEN, but not DON (Sydenham et al., 1991), and this capability was confirmed for several strains collected from Finland, Germany, Yugoslavia, Italy, Austria and Poland (Golinski et al., 1988; Bottalico et al., 1990). In 1988–89 surveys of Austrian *Fusarium*-infected ears, in spite of the wide presence of *F. graminearum*, Lew et al. (1991) ascribed the occurrence of NIV to the presence of *F. cerealis*. Moreover, in the 1988 Polish survey of maize red ear rot, almost all samples predominantly colonized by *F. cerealis* were found to be highly contaminated by NIV and FUS (Visconti et al., 1990). Other Polish surveys (1990–91) of maize ear rot highly infected with *F. graminearum* and *F. cerealis* found DON, 15-AcDON and ZEN in ears mostly colonized by *F. graminearum*, while high concentrations of NIV, FUS and ZEN were found in ears predominantly infected by *F. cerealis* (Grabarkiewicz-Szczesna et al., 1996). Finally, in Polish samples of maize ears affected by pink ear rot caused by *F. poae*, NIV was found, associated with FUS, both in grains and in cobs (Chelkowski et al., 1994a).

Occurrence of T-2 and HT-2 toxins

Epidemics of *F. sporotrichioides* usually lead to the accumulation of T-2 derivatives. Polish surveys of maize ears in 1984–85 showed *F. sporotrichioides* as the predominant fungus in approximately 2% of the samples. In hand-selected heavily damaged kernels, up to $1715\,mg\,kg^{-1}$ of total type A trichothecenes were found. These included T2, and HT2 (992 and $642\,mg\,kg^{-1}$, respectively), and lesser amounts of NOS, T-2 triol and T-2 tetraol (Chelkowski et al., 1987; 1989).

Occurrence of mono- and di-acetoxyscirpenol

Sporadic epidemics of *F. poae* can be responsible for the accumulation, in infected ears, of DAS and MAS, usually associated with NIV and FUS. In maize samples from Austria, NIV and FUS, and DON and ZEN, all probably produced by a NIV-chemotype of *F. graminearum* and *F. culmorum*, were found together with DAS presumably formed by *F. poae* and *F. cerealis* (Bottalico et al., 1983). In Poland, where *F. poae* is a widespread agent of maize pink ear rot, DAS, and MAS, together with NIV and FUS, were produced significantly by almost all the strains (11/14) collected from different locations (Chelkowski et al., 1994a).

Occurrence of fumonisins

There is increasing evidence of the occurrence of FB_1 in maize and maize-based food and feed all over the world (WHO, 2000), including Europe (Sanchis et al., 1994; Pestka et al., 1994; Doko and Vizconti, 1993; Visconti et al., 1996; Patel et al., 1997). Investigations carried out in Italy revealed that *F. verticillioides* was the most frequently isolated fungus from infected maize plants and from commercial maize kernels associated with FB_1 at levels of up to $5.31\,mg\,kg^{-1}$ (Pietri et al., 1995). In general, the occurrence of FB_1 appears to be important in southern European areas, that is Portugal, France (Le Bars and Le Bars, 1995; Dragoni et al., 1996), Spain (Rapior et al., 1993), Croatia (Jurjević et al., 1999), and Italy (Doko and Visconti, 1994; Visconti et al., 1996). Fumonisins are primarily formed in plants infected with *F. verticillioides* and *F. proliferatum*, when they are frequently found in pre-harvested maize ears. In Italy, high levels of fumonisins were often found ($250\,mg\,kg^{-1}$, in 6/6 selected samples) in association with more severe pink ear rot than that reported for other European countries, (Bottalico et al., 1995; Bottalico, 1998). On the contrary, FB_1 levels seem significantly lower in central to north-eastern European areas, including Austria ($15\,mg\,kg^{-1}$) (Lew et al., 1991; Krska et al., 1997), Switzerland (Pittet et al., 1992), Germany (Usleber et al., 1994), the Czech Republic (Ostrý and Ruprich, 1997), Croatia (Jurjević et al., 1999), and Poland (Chelkowski et al., 1994b). Surveys carried out in Croatia on freshly harvested samples of maize kernels collected in 1996–97 from 14 counties showed a scant presence of *F. verticillioides* (9.4%) and *F. subglutinans* (12.2%) associated with a micro-contamination of $FB_1 + FB_2$ ($11.66\,mg\,kg^{-1}$ in 93% of positive samples), together with a predominant occurrence of *Penicillium* spp. (98.5% of samples). However, it seems that in some more northern areas, under very favourable conditions, the fumonisins could reach significant levels. This was reported for the Slovak Republic where the level of fumonisins are usually recorded as negligible (Piecková and Jesenská, 1997). The 1998 season was characterized by severe epidemics of *F. verticillioides* (up to 100% of infected ears) reinforced by the occurrence of *F. proliferatum* (up to 94%). In particular, in pre-harvest infected maize ears, not only were high concentrations of FB_1 ($26.9\,mg\,kg^{-1}$) and FB_2 found ($6.3\,mg\,kg^{-1}$), but the formation of FUP was marked and traces of BEA were also found (Srobárová et al., 2000).

In southern Europe, *F. proliferatum*, which represents an additional FB_1 source, is frequently found with *F. verticillioides*. Logrieco et al. (1995) and Bottalico et al. (1995) reported that almost all strains of *F. proliferatum* collected in Italy produced FB_1. Moreover, selected maize ears mainly infected by *F. verticillioides* were found to be contaminated with FB_1, together with BEA and MON (Logrieco et al., 1995). In an additional survey, FB_1 was found in 9 out of 12 selected maize ears, mainly infected by *F. proliferatum* (Ritieni et al., 1997a). Therefore, as a result of the co-occurrence of both the principal fungal sources of FB_1, *F. verticillioides* and *F. proliferatum*, the probability of finding these carcinogenic toxins in maize is higher in southern than in central or northern Europe.

Occurrence of moniliformin

In infected maize ears, the main MON-producing *Fusarium* species are *F. subglutinans*, *F. proliferatum* and *F. avenaceum*. While *F. subglutinans* is widespread in central to northern Europe, and sometimes in part replaces *F. verticillioides*, *F. proliferatum* is increasingly reported from south to central European maize-growing areas as reinforcing the occurrence of *F. verticillioides* (Logrieco et al., 1995). In Italy, high levels of MON were found in maize ears with a widespread presence of *F. proliferatum* in association with *F. verticillioides* (Logrieco et al., 1995). In addition, MON was commonly found in maize ears infected with *F. subglutinans* from central and northern European countries, including Austria (Lew et al., 1991), and Poland (Kostechi et al., 1995; 1997). In particular, very high amounts of MON were found in all infected maize ears in Poland during years with severe epidemics of *F. subglutinans* (Lew et al., 1996). The high occurrence of MON in maize ear rot from Austria and Poland seemed to be related not only to the spread of *F. subglutinans*, but also to the frequent occurrence of *F. avenaceum* (Lew et al., 1991; 1996). In nature, *F. subglutinans* appears to be a higher MON producer than *F. avenaceum*, and levels up to $399.3\,mg\,kg^{-1}$ were associated with its occurrence in infected kernels from ear rot. This level compares with very much lower MON concentrations associated with the presence of *F. avenaceum* (Sharman et al., 1991). The toxigenicity of *F. proliferatum* seems comparable with that of *F. subglutinans* and Logrieco et al. (1995) reported strains of *F. proliferatum* were able to produce very large amounts of MON on autoclaved maize. However, the co-occurrence of *F. avenaceum* with *F. subglutinans* on maize in central–northern Europe increases the risk of MON accumulation in infected ears. Therefore, it appears that MON could be one of the more expected mycotoxins in maize ear rot in Europe (Schütt et al., 1998).

Occurrence of beauvericin

In maize ears infected with *F. verticillioides*, *F. subglutinans* and *F. proliferatum*, the occurrence of BEA was found in several European maize-growing areas together with the formation of FB_1 by *F. verticillioides* and *F. proliferatum*, and/or MON by *F. subglutinans* and *F. proliferatum*. In particular, BEA was reported in Italy (Moretti et al., 1994; Bottalico et al., 1995; Ritieni et al., 1997a), Poland (Logrieco et al., 1993; Kostechi et al., 1995), Austria (Krska et al., 1996), and the Slovak Republic (Srobarovà et al., 2000). However, it can be presumed that BEA is more widespread than recorded to date. In effect, BEA is produced not only by *F. proliferatum* and *F. subglutinans*, which are widespread from south–central to central–northern areas respectively (Krska et al., 1997; Logrieco, 2000), but also by several other *Fusarium* species occurring less frequently even in epidemic form, including *F. avenaceum*, *F. acuminatum*, *F. equiseti*, *F. poae* and *F. oxysporum* (Logrieco et al., 2000). It is worth noting that Austrian strains of *F. subglutinans* and *F. proliferatum* produced high levels of BEA on autoclaved maize (Krska et al., 1997). Similar results were obtained by Bottalico et al. (1995) with strains of *F. proliferatum* from Sardinia.

Occurrence of fusaproliferin

An important emerging toxicological problem seems to be connected with the occurrence of FUP. This novel fusariotoxin is produced, mostly together with BEA, by strains of *F. proliferatum* and *F. subglutinans* from maize and maize-based food and feed (Moretti et al., 1995; 1997; Logrieco et al., 1996). Since its isolation and characterization, FUP has been increasingly found in maize pink ear rot associated with *F. proliferatum*, especially in Italy (Ritieni et al., 1997a), and in the Slovak Republic (Srobarovà et al., 2000). In particular, 15 of 39 samples of pre-harvest maize ears collected throughout Italy during 1993–94, and predominantly infected by *F. proliferatum,* were contaminated with FUP, in association with FB_1 and BEA (Ritieni et al., 1997a).

Future perspectives

The many reports from Europe on maize contamination by mycotoxigenic *Fusarium* species show that the colonization of this important crop plant could be determined by organisms with a broad range of mycotoxin production and different environmental niches. Therefore, the evident biodiversity of these fungal species calls not only for a complex approach from plant pathologists, but requires particular attention to different toxicological problems related to the specific *Fusarium* contamination. The correct identification of the *Fusarium* species contaminating maize in different areas is evidence for a need not only for studying the levels of the interaction between *Fusarium* pathogens and maize, but also to obtain a precise picture of the toxicological risks related to the maize consumption by humans and animals. Developing modern, easy and fast tools for correct identification of toxigenic *Fusarium* species is therefore one of the future challenges for *Fusarium* workers. Moreover, as the profile of maize contaminants seems to be related to different environmental conditions, a better evaluation of each *Fusarium* species should be extremely useful for developing risk assessment models using data collected from a large range of sites in Europe with contrasting climatic conditions. Therefore, constant monitoring of maize throughout Europe in order to collect data on the *Fusarium* species and mycotoxins that are present in the maize is needed for establishing a European data base. Finally, the occurrence in maize of several mycotoxins with specific chemical traits and modes of action is a serious problem because of their additive and/or synergistic effects. Investigations of these possible effects and of the risks related to multiple mycotoxin contamination of maize will be a major goal for scientists involved in studies of *Fusarium* in the whole Europe.

Acknowledgments

This work was supported by COST-835 "Agriculturally Important Toxigenic Fungi" and by EU project (QLK1-CT-1999-01380).

References

Abbas HK, Shier WT, Seo JA, Lee YW and Musser SM (1998) Phytotoxicity and cytotoxicity of the fumonisin C and P series of mycotoxins from *Fusarium* spp. fungi. Toxicon 36: 2033–2037

Adler A, Law H, Edinger W, Brodacz W, Kiemdler E, Oberforster M and Hinterholzer J (2001) Fusaria in Austrian cereals – shifting of species and toxin contamination. In: Abstracts of International Symposium on "Bioactive fungal metabolites – Impact and exploitation, Swansea, Wales, 22–27 April 2001, p. 40

Arino A and Bullerman LB (1994) Fungal colonization of corn grown in Nebraska in relation to year, genotype and growing conditions. Journal of Food Protection 57: 1084–1087

Beardall GM and Miller JD (1994) Diseases in humans with mycotoxins as possible causes. In: Miller JD and Trenholm HL (eds) Mycotoxins in Grain. Compounds Other Than Aflatoxin (pp 487–539) Eagan press, St. Paul, Minnesota, USA

Bhat RV, Beedu SR, Ramakrishna Y and Munshi KL (1989) Outbreak of trichothecene mycotoxicosis associated with consumption of mould-damaged wheat in Kashmir Valley, India. Lancet i: 35–37

Bočarov-Stančić A, Škrinjar M, Mašic Z, Pavcov S and Gološin B (1997) Natural occurrence of *Fusarium* spp. and fusariotoxins in Yugoslav corn kernels. Cereal Research Communications 25: 581–582

Bottalico A (1979) On the occurrence of zearalenone in Italy. Mycopathologia 67: 119–121.

Bottalico A (1997) Toxigenic *Fusarium* species and their mycotoxins in pre-harvest cereals in Europe. Bulletin of the Institute for Comprehensive Agricultural Sciences Kinki University, Nara, Japan 5: 47–62

Bottalico A (1998) *Fusarium* diseases of cereals: Species complex and related mycotoxin profiles in Europe. Journal of Plant Pathology 80: 85–103

Bottalico A, Lerario P and Visconti A (1983) Mycotoxins occurring in *Fusarium*-infected maize ears in the field, in some European countries. In: Proceedings of the International Symposium on Mycotoxins (pp 375–382) Publication of the Science Department NIDOC, Cairo

Bottalico A and Logrieco A (1988) Osservazioni sulla fusariosi del Mais in Basilicata. II. Influenza di alcuni fattori colturali. Informatore fitopatologico 38: 55–58

Bottalico A, Visconti A, Logrieco A, Solfrizzo M and Mirocha CJ (1985) Occurrence of zearalenols (diastereomeric mixture) in corn stalk rot and their production by associated *Fusarium* species. Applied and Environmental Microbiology 49: 547–551

Bottalico A, Logrieco A and Ricci V (1986) Osservazioni sulla fusariosi del Mais in Basilicata. I. Incidenza della malattia e specie di *Fusarium* coinvolte. Informatore fitopatologico 36: 27–30

Bottalico A, Logrieco A and Visconti A (1989) *Fusarium* species and their mycotoxins in infected corn in Italy. Mycopathologia 107: 85–92

Bottalico A, Logrieco A and Visconti A (1990) Mycotoxins produced by *Fusarium crookwellense* Burgess, Nelson and Toussoun. Phytopathologia Mediterranea 29: 24–127

Bottalico A, Logrieco A, Ritieni A, Moretti A, Randazzo G and Corda P (1995) Beauvericin and fumonisin B_1 in preharvest *Fusarium moniliforme* maize ear rot in Sardinia. Food Additives and Contaminants 12: 599–607

Bullermann LB (1996) Occurrence of *Fusarium* and fumonisins on food grains and in foods. In: Jackson LS, De Vries JV and Bullerman LB (eds) Fumonisins in Food (pp 27–38) Plenum Press, New York

Caramelli M, Dondo A, Cantini Cortellazzi G, Visconti A, Minervini F, Doko MB and Guarda F (1993) Equine leukoencephalomalacia from fumonisin: First case in Italy. Ippologia 4: 49–56

Charmley LL, Trenholm HL, Prelusky DA and Rosenberg A (1995) Economic losses and decontamination. Natural Toxins 3: 199–203

Chelkowski J (1989) Mycotoxins associated with corn cob fusariosis. In: Chelkowki J (ed.) *Fusarium* – Mycotoxins, Taxonomy and Pathogenicity (pp 53–62) Elsevier, Amsterdam

Chelkowski J, Kwasna H, Zajkowki P, Visconti A and Bottalico A (1987)*Fusarium sporotrichioides* Sherb. and trichothecenes associated with *Fusarium*-ear rot of corn before harvest. Mycotoxin Research 3: 111–114

Chelkowski J, Zawadzki M, Zajkowki P, Logrieco A and Bottalico A (1990) Moniliformin production by *Fusarium* species. Mycotoxin Research 6: 41–45

Chelkowski J, Lew H and Pettersson H (1994a) *Fusarium poae* (Peck) Wollenw. – Occurrence in maize ears, nivalenol production and mycotoxin accumulation in cobs. Mycotoxin Research 10: 116–120

Chelkowski J, Pronczuk M, Visconti A, Doko MB (1994b) Fumonisin B_1 and B_2 accumulation in maize kernels inoculated under field conditions with *Fusarium moniliforme* Sheldon and in naturally infected cobs in Poland. Genetica Polonica, 35B: 333–338

Ciudin E and Bazgan O (1991) Chromatographical and biological tests of some samples of maize grains with *F. graminearum* and *F. culmorum*. Cercetâri Agronomice în Moldova 24: 109–112

Doko MB and Visconti A (1993) Fumonisin contamination of maize and maize-based foods in Italy. In: Scudamore KA (ed.) Occurrence and Significance of Mycotoxins (pp 49–55) Central Science Laboratory, London, UK

Doko MB, Rapior S, Visconti A and Schjoth JE (1995) Incidence and levels of fumonisin contamination in maize genotypes grown in Europe and Africa. Journal of Agricultural and Food Chemistry 43: 429–434

Doko MB and Visconti A (1994) Occurrence of fumonisin B_1 and B_2 in corn and corn-based human foodstuffs in Italy. Food Additives and Contaminants 11: 433–439

Dragoni I, Piantanida L, Tirilly Y, Visconti A and Pascale M (1996) Occurrence of fumonisins in corn feedstuffs intended for pig consumption in Bretain (France). In: Abstract Book of IX International IUPAC Symposium on Mycotoxins and Phycotoxins, Rome, May 27–31, 1996, p. 39

Ehling G, Cockburn A, Snowdon P and Buschhaus H (1997) The significance of the *Fusarium* toxin deoxynivalenol (DON) for human and animal health. Cereal Research Communications 25: 443–447

Ellend N, Binder J, Krska R and Horvath EM (1997) Contamination of Austrian corn with *Fusarium* toxins in autumn 1996. Cereal Research Communications 25: 359–360

Eriksen GS and Alexander J (1998) *Fusarium* Toxins in Cereals – a Risk Assessment. TemaNord 1998:502. Nordic Council of Ministers. Ekspressen Tryk and Kopicenter, Copenaghen, Denmark, 146 pp

Franceschi S, Bidoli E, Baron AE and La Vecchia C (1990) Maize and the risk of cancer of the oral cavity, pharynx and oesophagus in North-Eastern Italy. Journal of National Cancer Institute 82: 1407–1411

Gelderblom WCA, Snyman SD, Abel S, Lebepe-Mazur S, Smuts CM, Van der Westhuizen L, Marasas WFO, Victor TC, Knasmuller S and Huber W (1996) Hepatotoxicity and carcinogenicity of the fumonisins in rats. A review regarding mechanistic implications for establishing risk in humans. In: Jackson LS, De Vries JW and Bullerman LB (eds) Fumonisins in Food (pp 279–296) Plenum Press, New York

Golinski P, Vesonder RF, Latus-Zietkiewicz D and Perkowski J (1988) Formation of fusarenone X, nivalenol, zearalenone, α-trans-zearalenol, β-trans-zearalenol, and fusarin C by *Fusarium crookwellense*. Applied and Environmental Microbiology 54: 2147–2148

Grabarkiewicz-Szczesna J, Foremska E and Golinski P (1996) Distribution of trichothecene mycotoxins in maize ears infected with *F. graminearum* and *F. crookwellense*. Mycotoxin Research 12: 45–50

Gupta S, Krasnoff SB, Underwood NL, Renwick JAA and Roberts DW (1991) Isolation of beauvericin as an insect toxin from *Fusarium semitectum* and *Fusarium moniliforme* var. *subglutinans*. Mycopathologia 115: 185–189

Harrison LR, Colvin BM, Greene JT, Newman LE and Cole JR (1990) Pulmonary edema and hydrotorax in swine produced by fumonisin B_1, a toxic metabolite of *Fusarium moniliforme*. Journal Veterinary Diagnosis Investigations 2: 217–221

Harvey RB, Kubena RF, Rottinghaus GE, Turk JR and Buckley SA (1997) Effects of fumonisin and moniliformin from culture materials to growing swine. Cereal Research Communications 25: 415–417

Hopmans EC and Murphy PA (1995) Fumonisins: Mycotoxins produced by *Fusarium* species. In: Bidlack WR and Omaye ST (eds) Natural Protectants Against Natural Toxicants. Vol. I. Natural Protectants and Natural Toxicants in Food (pp 61–78) Technomic Pub. Co. Inc., Lancaster – Basel

IARC (1993) IARC Monographs on the Evaluation of Carcinogenic risks to Humans. Vol. 56. Some Naturally Occurring Substances: Food Items and Constituents, Heterocyclic Aromatic Amines and Mycotoxins. International Agency for Research on Cancer, Lyon, 397–444; 445–466; 467–488

Javed T, Bennett GA, Richard JL, Dombrink-Kurtzman MA, Coté LM and Buck WB (1993): Mortality in broiler chicks on feed amended with *Fusarium proliferatum* culture material or with purified fumonisin B_1, and moniliformin. Mycopathologia 123: 171–184

Jurjević Z, Cvjetković B, Jurjević and Ceović V (1997) Occurrence of *Fusarium* spp. on corn in the region of endemic nephropathy in Croatia. Cereal Research Communications 25: 455–456

Jurjević Z, Solfrizzo M, Cvjetković B, Avantaggiato G and Visconti A (1999) Ochratoxin A and fumonisins (B_1 and B_2) in maize from Balkan nephropathy endemic and non endemic areas of Croatia. Mycotoxin Research 15: 67–80

Kellerman TS, Marasas WFO, Thiel PG, Gelderblom WGA, Cawood M and Coetzer JAW, (1990) Leukoencephalomalacia in two horses induced by oral dosing of fumonisin B_1. Onderstepport Journal of Veterinary Research 2: 269–275

Kostechi M, Szczesna J, Chelkowski J and Wisniewska H (1995) Beauvericin and moniliformin production by Polish isolates of *Fusarium subglutinans* and natural co-occurrence of both mycotoxins in cereal grain samples. Microbiologie, Aliments, Nutrition 13: 67–70

Kostechi M, Grabarkiewicz-Szczesna J and Golinski P (1997) Simultaneous analysis of beauvericin and moniliformin in fungal cultures and in cereal grain samples. Mycotoxin Research 13: 17–22

Krska R, Lemmens M, Schuhmacher R, Grasserbauer M, Pronczuk M, Wisniewska H and Chelkowski J (1996) Accumulation of the mycotoxin beauvericin in kernels of corn hybrids inoculated with *Fusarium subglutinans*. Journal of Agricultural and Food Chemistry 44: 3665–3667

Krska R, Schuhmacher R, Grasserbauer M, Lemmens M, Lemmens-Gruber R, Adler A and Lew H (1997) Effects of beauvericin to mammalian tissue and its production by Austrian isolates of *Fusarium proliferatum* and *Fusarium subglutinans*. Mycotoxin Research, 13: 11–16

Krüger W, (1989) Maize diseases caused by *Fusaria*: Involved species and mycotoxins. In: Chelkowki J (ed.) *Fusarium* – Mycotoxins, Taxonomy and Pathogenicity (pp 297–317) Elsevier, Amsterdam

Kuiper-Goodman T, Scott PM and Watanabe H (1987) Risk assessment of the mycotoxin zearalenone. Regulatory Toxicology and Pharmacology 7: 253–259

Le Bars P and Le Bars J (1995) Ecotoxinogenesis of *Fusarium moniliforme*: Appearance of risks of fumonisins. Cryptogamie, Mycologie 16: 59–64.

Ledoux DR, Bermudez AJ, and Rottinghaus GE (1995) Effects of feeding *Fusarium fujikuroi* culture material, containing known levels of moniliformin, in young broiler chicks. Poultry Science 74: 297–305

Leslie JF (1995) *Gibberella fujikuroi*: Available populations and variable traits. Canadian Journal of Botany 73 (Suppl 1): S282–S291

Lević J, Tamburić-Ilinčić L and Petrović T (1997) Maize kernel infection by *Fusaria* in the period 1994–1996. Cereal Research Communications, 25: 773–775

Lew H, Adler A and Edinger W (1991) Moniliformin and the European corn borer (*Ostrinia nubilalis*). Mycotoxin Research 7A: 71–76

Lew H, Chelkowski J, Pronczuk P and Edinger W (1996) Occurrence of the mycotoxin moniliformin in maize (*Zea mays* L.) ears infected by *Fusarium subglutinans.* Food Additives and Contaminants 13: 321–324

Lew H, Adler A, Edinger W (1997) Dynamics of the *Fusarium* toxin distribution in maize plants affected by stalk rot. Cereal Research Communications 25: 467–470

Logrieco A (2000) Advances on natural occurrence and production of beauvericin by *Fusarium* species. In: Abstracts of 6th European *Fusarium* Seminar (pp 42–43) 11–16 September, Berlin

Logrieco A, Altomare C, Mulè G and A Bottalico (1992) Alcuni dati sulla presenza e patogenicità di chemiotipi di *Fusarium graminearum* in Europa. Atti Giornate Fitopatologiche 2, 287–294

Logrieco A, Moretti A, Ritieni A, Chelkowski J, Altomare A, Bottalico A and Randazzo G (1993) Natural occurrence of beauvericin in pre-harvest *Fusarium subglutinans* infected corn ears in Poland. Journal of Agricultural and Food Chemistry 41: 2149–2152

Logrieco A, Moretti A, Ritieni A, Bottalico A and Corda P (1995) Occurrence and toxigenicity of *Fusarium proliferatum* from pre-harvest maize ear rot, and associated mycotoxins, in Italy. Plant Disease 79: 727–731

Logrieco A, Moretti A, Fornelli F, Fogliano V, Ritieni A, Caiaffa MF, Randazzo G, Bottalico A and Macchia L (1996) Fusaproliferin production by *Fusarium subglutinans* and its toxicity to *Artemia salina*, SF-9 insect cells, and IARC/LCL 171 human B lymphocytes. Applied and Environmental Microbiology 62: 3378–3384

Macchia L, Di Paola R, Fornelli F, Nenna S, Moretti A, Napolitano R, Logrieco A, Caiaffa MF, Tursi A and Bottalico A (1995) Cytotoxicity of beauvericin to mammalian cells. In: Abstracts of International Seminar on '*Fusarium*: Mycotoxins, Taxonomy and Pathogenicity' May 9–13 (pp 72–73) Martina Franca,

Marasas WFO (1995) Fumonisins: Their implications for human and animal health. Natural Toxins 3: 193–198

Marasas WFO, Thiel PG, Rabie CJ, Nelson PE and Toussoun TA (1986) Moniliformin production in *Fusarium* section *Liseola*. Mycologia 78: 242–247

Milevoj L (1997) Electrophoretic study of proteins in the fungus *Fusarium moniliforme* var. *subglutinans*. Cereal Research Communications 25: 603–606

Moretti A, Logrieco A, Bottalico A, Ritieni A and Randazzo G (1994) Production of beauvericin by *Fusarium proliferatum* from maize in Italy. Mycotoxin Research 10: 73–78

Moretti A, Logrieco A, Bottalico A, Ritieni A, Randazzo G and Corda P (1995) Beauvericin production by *Fusarium subglutinans* from different maize geographical areas. Mycological Research 99: 282–286

Moretti A, Logrieco A, Bottalico A, Ritieni A, Fogliano V and Randazzo G (1997) Diversity in beauvericin and fusaproliferin production by different populations of *Gibberella fujikuroi* (*Fusarium* section *Liseola*). Sydowia, 48: 44–56

Munkvold GP, McGee DC and Carlton WM (1997) Importance of different pathways for maize kernek infection by *Fusarium moniliforme*. Phytopathology 87: 209–217

Musser SM, Gay ML, Mazzola EP and Plattner RD (1996) Identification of a new series of fumonisins containing 3-hydroxypyridine. Journal of Natural Products 59: 970–972

Nagy E and Bâgiu L (2000) Aggressiveness of *Fusarium graminearum* and *Fusarium moniliforme* in maize. In: Abstracts of the 6th European *Fusarium* Seminar (pp 80–81) 11–16 September, Berlin

Nedelnik J (2000) Spectrum of *Fusarium* species and contamination with mycotoxins of corn in the Czech Republic. In: Abstracts of the 6th European *Fusarium* Seminar (pp 111–112) 11–16 September, Berlin

Nelson PE, Toussoun TA and Marasas WFO (1983) Fusarium Species – An Illustrated Manual for Identification, Pennsylvania State University Press, University Park, Pennsylvania, USA

Nelson PE, Plattner RD, Shackelford DD and Desjardins AE (1992) Fumonisin B_1 production by *Fusarium* species other than *F. moniliforme* in section *Liseola* and by some related species. Applied and Environmental Microbiology 58: 984–989

Oldenburg E (1993) Occurrence of zearalenone in maize. Mycotoxin Research 9: 72–78

Ostrý V and J Ruprich (1997) Fumonisiny, mykotoxiny produkované druhy rodu *Fusarium*. *Mykologické Listy*, 60: 11–18 (Reported by Review of Plant Pathology, 1997, 76:1147. Abstract No. 8853)

Osweiler G (1995) Fumonisins and pulmonary edema in swine. In: Bidlack WR and Omaye ST (eds) Natural Protectants against Natural Toxicants. Vol. I, Natural Protectants and Natural Toxicants in Food (pp 79–87) Technomic Publishing Corporation, Lancaster – Basel

Pascale M, Doko MB and Visconti A (1995) Determination of fumonisins in polenta by high performance liquid chromatography. In: Proceedings of the 2nd National Congress on Food Chemistry, 24–27 May (pp 1067–1071) Giardini Naxos, Messina

Patel S, Hazel CM, Winterton AGM and Gleadle AE (1997) Surveillance of fumonisins in UK maize-based foods and other cereals. Food Additives and Contaminants 14: 187–191

Perkowski J, Chelkowski J, Plattner RD and Golinski P (1991) Cumulation of mycotoxins in maize cobs infected with *Fusarium graminearum*. Mycotoxin Research 7: 115–120

Pestka J, Azcona-Olivera JI, Plattner RD, Minervini F, Doko MB and Visconti A (1994) A comparative assessment of fumonisin in grain-based foods by ELISA, GC-MS and HPLC. Journal of Food Protection 57: 169–172

Piecková E and Jesenká Z (1997) *Fusarium moniliforme* and *F. subglutinans* in maize-based foodstuffs in the Slovak Republic. Cereal Research Communications 25: 609–610

Pietri A, Bertuzzi T and Piva G (1995) Fumonisin contamination of maize grown in Italy. In: Abstracts of International Seminar on '*Fusarium*: Mycotoxins, Taxonomy and Pathogenicity', May 9–13 (pp 18–19) Martina Franca

Pittet A, Parisod V and Schellenberg M (1992) Occurrence of fumonisins B_1 and B_2 in corn-based products from the Swiss market. Journal of Agricultural and Food Chemistry 40: 1352–1354

Plattner RD and Nelson PE (1994) Production of beauvericin by a strain of *Fusarium proliferatum* isolated from corn fodder for swine. Applied and Environmental Microbiology 60: 3894–3896

Ramakrishnan NY and Wu WD (1994) Toxicity of corn culture material of *Fusarium proliferatum* M-7176 and nutritional intervention in chicks. Poultry Science 73: 617–626

Rapior S, Miller JD, Savard ME and ApSimon JW (1993) Fumonisin and fusarin production *in vitro* by European strains of *Fusarium moniliforme*. Microbiologie, Aliments, Nutrition 11: 327–333

Rheeder JP, WFO Marasas, PG Thiel, Sydenham EW, Shephard GS and van Schalkwyk DJ (1992) *Fusarium moniliforme* and fumonisins in corn in relation to human oesophageal cancer in Transkei. Phytopathology 82: 353–357

Ritieni A., Fogliano V, Randazzo G, Scarallo A, Logrieco A, Moretti A, Bottalico A and Mannina L (1995) Isolation and characterization of fusaproliferin, a new toxic metabolite from *Fusarium proliferatum*. Natural Toxins 3: 17–20

Ritieni A, Moretti A, Logrieco A, Bottalico A, Randazzo G, Monti MS, Ferracane R and Fogliano V (1997a) Occurrence of fusaproliferin, fumonisin B_1, and beauvericin in maize from Italy. Journal Agricultural and Food Chemistry 45: 4011–4016

Ritieni A, Monti MS, Randazzo G, Logrieco A, Moretti A, Peluso G, Ferracane R and Fogliano V (1997b) Teratogenic effects of fusaproliferin on chicken embryos. Journal of Agricultural and Food Chemistry 45: 3039–3043

Rizzo A, Atroshi F, Hirvi T and Saloniemi H (1992) The hemolytic activity of deoxynivalenol and T-2 toxin. Natural Toxins 1: 106–110

Ross PF, Nelson PE, Richard JL, Osweiler GD, Rice LG, Plattner RD and Wilson TM (1990) Production of fumonisins by *Fusarium moniliforme* and *F. proliferatum* associated with equine leukoencephalomalacia and a pulmonary edema syndrome in swine. Applied and Environmental Microbiology 56: 3225–3226

Rotter BA and Prelusky DB (1996) Toxicology of deoxynivalenol (vomitoxin). Journal of Toxicology and Environmental Health 48: 1–34

Sanchis V, Abadias M, Oncins L, Sala N, Vinas I and Canela R (1994) Occurrence of fumonisins B_1 and B_2 in corn-based products from the Spanish market. Applied and Environmental Microbiology 60: 2147–2148

Sanchis V, Abadias M, Oncins L, Sala N, Viñas I and Canela R (1995) Fumonisins B_1 and B_2 and toxigenic *Fusarium* strains in feeds from the Spanish market. International Journal of Food Microbiology 27: 37–44

Schütt F, Nirenberg HI and Deml G (1998) Moniliformin production in the genus *Fusarium*. Mycotoxin Research 14: 35–40

Sharman M, Gilbert J and Chelkowski J (1991) A survey on the occurrence of moniliformin in cereal samples from sources worldwide. Food Additives and Contaminants 8: 459–466

Shephard GS, Thiel PG, S Stockenström and Sydenham EW (1996) Worldwide survey of fumonisin contamination of corn and corn-based products. Journal of the Association of Official Analytical Chemists International 79: 671–687

Shurtleff MC (1980) Compendium of Corn Diseases. The American Phytopathological Society, 128 pp

Srobárová A (1997) Occurrence of *Fusarium* spp. from Slovakia maize kernels. Cereal Research Communications 25: 617–618

Srobárová A., Logrieco A, Ritieni A, Ferracane A, Moretti A and Nadubinská M (2000) Contamination of the maize ear by toxigenic *Fusarium* spp. In Slovakia. In: Abstracts of the 6th European *Fusarium* Seminar, 11–16 September (pp 44–45) Berlin

Sydenham EW, Marasas WFO, Thiel PG, Shephard GS and Nieuwenhuis JJ (1991) Production of mycotoxins by selected *Fusarium graminearum* and *F. crookwellense* isolates. Food Additives and Contaminants 8: 31–41

Szécsi A (1994) Occurrence of *Fusarium* species in section *Liseola* isolated from Hungarian corn samples in 1991 and 1992. Novenyvedelem 30: 313–318

Szécsi A and Bartok T (1995) Trichothecene chemotypes of *Fusarium graminearum* from corn in Hungary. Mycotoxin Research 11: 85–92

Usleber E, Straka M and Terplan G (1994) Enzyme immunoassay for fumonisin B_1 applied to corn-based food. Journal Agricultural Food Chemistry 42: 1392–1396

Visconti A (1996) Fumonisins in maize genotypes grown in various geographic areas. In: Jackson LS, De Vries WJ and Bullerman LB (eds) Fumonisins in Food (pp 193–204) Plenum Press, New York

Visconti A, Chelkowski J, Solfrizzo M and Bottalico A (1990) Mycotoxins in corn ears naturally infected with *Fusarium graminearum* and *F. crookwellense*. Canadian Journal of Plant Pathology 12: 187–189

Visconti A, Pascale M and Doko MB (1996) Fumonisins in Europe. Production by *Fusarium* isolated from cereals and occurrence in maize and maize products intended for human and animal consumption. In: Marengo G and Pastoni F (eds) The European Union's Innovative Policy Against Food-Transmitted Disease (pp 162–174) EUR 16414 EN, ECSC-EC-EAEC, Brussels, Luxembourg,

Wang ZG, Feng J and Tong Z (1993) Human toxicosis caused by moldy rice contaminated with *Fusarium* and T-2 toxin. Biomedical Environmental Science 6: 65–70

Warfield CY and Gilchrist DG (1999) Influence of kernel age on fumonisin B_1 production in maize by *Fusarium moniliforme*. Applied and Environmental Microbiology 65: 2853–2856

WHO (2000) International Programme on Chemical Safety. Environmental Health Criteria 219: Fumonisin B_1. World Health Organization, Geneva, 150 pp

European Journal of Plant Pathology **108:** 611–624, 2002.

Toxigenic *Fusarium* species and mycotoxins associated with head blight in small-grain cereals in Europe

Antonio Bottalico and Giancarlo Perrone
Institute of Sciences of Food Productions (ISPA), National Council of Researches, Viale Luigi Einaudi 51, 70125, Bari, Italy (Phone: +390805912860; Fax: +390805486063; E-mail: itmpab11@area.ba.cnr.it)

Key words: Fusarium head blight, *Fusarium graminearum*, *Fusarium culmorum*, *Fusarium avenaceum*, *Fusarium*-mycotoxins, deoxynivalenol, nivalenol, moniliformin, beauvericin, wheat

Abstract

The *Fusarium* species predominantly found associated with *Fusarium* head blight (FHB) in wheat and other small-grain cereals all over Europe are *F. graminearum*, *F. avenaceum* and *F. culmorum*. Among the less frequently encountered species are several others which are less pathogenic or opportunistic, but also toxigenic. These include *F. poae*, *F. cerealis*, *F. equiseti*, *F. sporotrichioides*, *F. tricinctum* and, to a lesser extent, *F. acuminatum*, *F. subglutinans*, *F. solani*, *F. oxysporum*, *F. verticillioides*, *F. semitectum* and *F. proliferatum*. The species profile of FHB is due to several factors, primarily climatic conditions, particularly rain and the temperature at flowering stage, but also agronomic factors, such as soil cultivation, nitrogen fertilization, fungicides, crop rotation, and host genotype. The most frequently encountered *Fusarium* mycotoxins in FHB in Europe has proved to be deoxynivalenol and zearalenone produced by *F. graminearum* and *F. culmorum*, with the former more common in southern (warmer) and the latter in northern (colder) European areas. Nivalenol was usually found associated with deoxynivalenol and its derivatives (mono-acetyldeoxynivalenols), together with fusarenone-X, formed by *F. graminearum*, *F. cerealis*, *F. culmorum* and, in northern areas, by *F. poae*. Moreover, from central to northern European countries, moniliformin has been consistently reported, as a consequence of the widespread distribution of *F. avenaceum*, whereas the occurrence of T-2 toxin derivatives, such as T-2 toxin and HT-2 toxin, and diacetoxyscirpenol have been recorded in conjunction with sporadic epidemics of *F. sporotrichioides* and *F. poae*. Finally, beauvericin and various enniatins have recently been found in Finnish wheat colonized by *F. avenaceum* and *F. poae*.

Abbreviations: AcDON = Mono-acetyldeoxynivalenols (3-AcDON, 15-AcDON); AcNIV = Mono-acetylnivalenol (15-AcNIV); BEA = Beauvericin; DiAcDON = Di-acetyldeoxynivalenol (3,15-AcDON); DAcNIV = Diacetylnivalenol (4,15-AcNIV); DAS = Diacetoxyscirpenol; DON = Deoxynivalenol (Vomitoxin); ENS = Enniatins; FUS = Fusarenone-X (= 4-Acetyl-NIV); HT2 = HT-2 toxin; MAS = Monoacetoxyscirpenol; MON = Moniliformin; NEO = Neosolaniol; NIV = Nivalenol; T2 = T-2 toxin; T2ol = T-2 Tetraol; ZEN = Zearalenone; ZOH = zearalenols (α and β isomers).

Introduction

Several *Fusarium* species are widespread pathogens on small-grain cereals (soft and durum wheat, barley, oats, rye and triticale) around the world, including all European cereal-growing areas. They can cause root, stem and ear rot, resulting in severe reductions in crop yield, often estimated at between 10% and 40%. In addition, several *Fusarium* strains are capable of producing mycotoxins which can be formed in pre-harvest infected plants still standing in the fields, or in stored grain (Bottalico, 1998). Mycotoxins in wheat and barley, which constitute almost two-thirds of the world production of small-grain cereals and almost 80% of the European small-grain production, are causing great concern, because of the extent of infection and contamination of food products. However, the other less frequently-grown small grains have also been reported to contain *Fusarium* mycotoxins, though these crops appear, in general, to be less susceptible to *Fusarium* head blight (FHB) and consequently less at risk from pre-harvest toxin contamination.

The occurrence of mycotoxins in small cereal grains, particularly in wheat, is of great concern worldwide, because their presence in processed feeds and foods seems unavoidable. Consequently, they have been associated with chronic or acute mycotoxicoses in livestock and, to a lesser extent, in humans (IARC, 1993). In this respect, it appears that almost all wheat and barley crops grown in northern European areas are to some extent contaminated, at least by deoxynivalenol (DON) whereas the southern areas seem much less contaminated by *Fusarium* toxins (Bottalico, 1998; Eriksen and Alexander, 1998; Visconti, 2001). The numerous books and reviews recently published demonstrate the increasing interest in FHB and related *Fusarium* toxins in cereal grains. These deal with the several biological aspects of the causal *Fusarium* species, and with the natural occurrence, toxicology and risk assessment of toxic compounds (Chełkowski, 1989a; 1991a; IARC, 1993; Miller and Trenholm, 1994; Miller, 1995; Bottalico, 1997). In addition, several FHB problems have been discussed and published during the six European Seminars so far dedicated to *Fusarium* taxonomy, pathogenicity and mycotoxins (Chełkowski, 1989a; 1991b; Mesterházy, 1997a; Logrieco et al., 1997) as well as within joint European Fusarium projects (Nirenberg, 1995a). Moreover, it is noteworthy that specific Committees operate in several European countries on *ad hoc* programmes dealing with several aspects of FHB in small-grain cereals. Among their aims, is the improvement of the resistance to the disease and the prevention of toxin accumulation (Eriksen and Alexander, 1998; LBP, 2000; Pasquini et al., 2001).

In this paper, the most relevant aspects of the distribution of toxigenic *Fusarium* species and related mycotoxins in ear rot of small-grain cereals in European countries are reviewed, with emphasis on the occurrence of DON, nivalenol (NIV) and other trichothecene derivatives, zearalenone (ZEN), and moniliformin (MON) in infected plants standing in the field, or in freshly harvested grain. The relationship between the *Fusarium* species complex and the relative mycotoxin profile, besides underlining the toxicological risk of FHB, may be useful for predicting the mycotoxins that are most likely to be formed in infected kernels, with reference to the most conducive factors, such as predominant *Fusarium* species, geographical area, environmental conditions, agronomic factors, and host genotype. It has been reported that in organic farming systems, which employ particular safety practices, the incidence of *Fusarium* spp. appeared comparatively lower than that in conventional systems. The DON or ZEN contents in freshly harvested wheat kernels were the same, but this aspect still a critical controversial point.

Fusarium species involved and mycotoxins produced

Fusarium *head blight of small-grain cereals*

Fusarium species which are pathogenic on wheat, barley and other small-grain cereals, are responsible for two forms of disease; 'foot rot' affects roots and crowns, and includes an early stage causing seedling blight and FHB, affecting individual kernels, single ear spikelets or entire heads, leading to scab of the kernels. Infected spikelets first appear water-soaked, lose their chlorophyll and become straw-coloured. In warm, humid weather, pinkish-red mycelium and conidia develop abundantly in the infected spikelets, and the infection spreads to adjacent spikelets or through the entire head. Infected kernels become shrivelled and discoloured with a white, pink, or light brown scaly appearance as a result of the mycelial outgrowths from the pericarp. From the mycotoxicological point of view, the phase of *Fusarium* diseases of small-grain cereals which is of greatest concern is FHB because of the potential accumulation of mycotoxins in grain intended for foods and feeds.

The risks connected with the consumption of contaminated forage and straw by livestock must not be ignored. FHB causes severe damage to wheat and the other cereals, especially in areas with warm temperature and high relative humidity during the heading and flowering period. Together with favourable environmental conditions, other factors determining the severity of the disease include various agronomic factors and host genotype (LBP, 2000).

Fusarium species involved

The *Fusarium* species commonly isolated from small-grain cereals in Europe are illustrated in Table 1. Their ability to synthesize mycotoxins was first illustrated by Bottalico (1998). The aetiological characteristic of FHB is the co-occurrence or the quick succession of several species of *Fusarium* usually referred to as a 'complex'. In fact, it is quite common to isolate up to nine different *Fusarium* species from a single fragment of infected tissue, or up to seventeen different species from freshly-harvested grain samples collected in a limited area. However, only a small number of species have been regarded as pathogenic and generally, very few of them predominate in a particular host-agroclimatic system. Similar to strains of the pathogenic and predominant *Fusarium* species, several other strains of the less pathogenic or opportunistic *Fusarium* species are capable of producing considerable amounts of mycotoxins. Therefore, the mycotoxigenic profile of a contaminated crop is due not only to the predominant pathogenic *Fusarium* species, but also to the opportunistic species making up the complex. Moreover, the host genotype may play an important role in toxin accumulation. In general, less toxins are formed in the more resistant genotypes (Mesterházy et al., 1991; Mesterházy, 1997b), or in less visibly moulded ears. However, cultivars of wheat showing low FHB severity have been reported with high DON content, and wheat cultivars highly susceptible to FHB may contain lower DON levels. In general, a direct prediction of toxin contamination based on the incidence of FHB, crop yield loss, or scabby grain severity cannot be performed accurately, especially when *Microdochium nivale* or *F. avenaceum* and other *Fusarium* species which do not produce DON predominate in the FHB complex.

The species predominantly found associated with head blight of wheat and other small-grain cereals are *F. graminearum* (and its widespread teleomorph *Gibberella zeae*), *F. culmorum*, and *F. avenaceum* (*G. avenacea*). Among the other less frequently isolated species are *F. poae*, *F. cerealis* (syn. *F. crookwellense*), *F. equiseti* (syn. *F. scirpi*) (*G. intricans*), *F. sporotrichioides*, and *F. tricinctum*. Many other species may be sporadically encountered, including *F. acuminatum*, *F. subglutinans* (syn. *F. sacchari*), *F. solani*, *F. oxysporum*, *F. semitectum* (syn. *F. pallidoroseum*, *F. incarnatum*), *F. verticillioides* (syn.: *F. moniliforme*), and *F. proliferatum*.

Recent mycological, toxicological and molecular studies have clarified many peculiar characteristics of the most important FHB agents, some of which are briefly described. Two populations were previously characterized within *F. graminearum* (*G. zeae*), designated as Group 1 and Group 2. These were differentiated because of colony growth rates and colony morphology on potato dextrose agar (PDA) slants, but they had almost the same toxigenic potential. Group 1 rarely formed perithecia in nature and mainly caused crown rot of cereals and

Table 1. Mycotoxigenic species isolated from FHB of wheat in Europe

Species	Geographical incidence		Mycotoxin
	North/Centre	South	
F. graminearum	+ + +	+ + +	**DON**, **NIV**, **ZEN**, AcDON, FUS
F. avenaceum	+ + +	++	**MON**, BEA, ENS
F. culmorum	+ + +	++	**DON**, **ZEN**, ZOH,NIV
F. poae	++	+	**NIV**, **BEA**, DAS, FUS, ENS
F. equiseti	++	+	**DAS**, ZEN, ZOH
F. tricinctum	+	+	**MON**
F. cerealis	+	±	**NIV**, **FUS**, **ZEN**, ZOH
F. sporotrichioides	+	±	**T2**, **HT2**, **T2ol**, NEO
F. acuminatum	±	±	**T2**, NEO
F. subglutinans	±	—	**MON**
F. solani	±	—	—
F. oxysporum	±	—	—

AcDON = Monoacetyl-deoxynivalenols (3-AcDON, 15-AcDON); BEA = Beauvericin; DAS = Diacetoxyscirpenol; DON = Deoxynivalenol (Vomitoxin); ENS = Enniatins; FUS = Fusarenone-X (4-Acetyl-NIV); HT2 = HT-2 toxin; MON = Moniliformin; NEO = Neosolaniol; NIV = Nivalenol; T2 = T-2 toxin; T2ol = T-2 tetraol; ZEN = Zearalenone; ZOH = zearalenols (α and β isomers).

grasses, while Group 2 formed abundant perithecia in nature and mainly caused FHB in small-grain cereals (Burgess et al., 1997). Recent investigations on anamorphic and teleomorphic characteristics of the two *F. graminearum* groups, as well as on their phylogenetic relationships and on different modes of sexual reproduction, have led Aoki and O'Donnell (1999a,b) to consider *F. graminearum* Group 1 as a separate species. It was named *F. pseudograminearum* Aoki et O'Donnell. Its teleomorph was characterized through mating experiments and named *Gibberella coronicola* Aoki et O'Donnell. The existence of distinct genetic groups within several strains of *F. graminearum* collected from the USA, Northwest Europe and Nepal has also been reported. These show different pathogenic and toxigenic capabilities on wheat and maize. Moreover, from a toxicological point of view, the toxigenic strains of *F. graminearum* were classified into DON and NIV chemotypes, with NIV chemotypes being strains mainly associated with maize (Sydenham et al., 1991; Szécsi and Bartok, 1995; Yoshizawa, 1997). Furthermore, DON chemotype strains of *F. graminearum* were sub-classified into two types: DON-chemotype IA producing DON and 3-AcDON, essentially from warmer regions (mostly including the European strains), and DON-chemotype IB, producing DON and 15-AcDON from slightly cooler regions (mostly including the American strains) (Miller et al., 1991; Wang and Miller, 1994). Finally, it seems that DON and NIV chemotypes of *F. graminearum* are not distributed evenly around the climatic areas of the world, and such ecological differences in chemotype distribution may contribute to establishing specific regional grain contamination. However, some factors of variability, including a drastic change in climatic conditions at a regional as well as at a continental level can, along with close crop rotations between maize and small-grain cereals, lead to changes in the spread of *Fusarium* species.

Toxigenic strains of *F. culmorum* have been split into two types in relation to the main type B trichothecenes produced, that is DON and NIV chemotypes. Strains of DON-types also produced AcDON (3-AcDON) (Miller et al., 1991; Gang et al., 1998; D'Mello et al., 1997), while strains of NIV-type were also able to produce FUS. In field trials, DON and NIV chemotypes exhibited different aggressiveness toward winter rye (Gang et al., 1998).

Within the section *Sporotrichiella*, a *Fusarium* species described as 'powdery *F. poae*' and responsible for the occurrence of T-2 toxin and its deacetylated derivative HT-2 toxin in Norwegian cereals, has recently been reported. Powdery *F. poae*, which was routinely identified as *F. poae* appears morphologically similar to *F. poae*, but with a mycotoxin profile similar to that of *F. sporotrichioides*. Preliminary results of molecular sequencing indicated that all strains of powdery *F. poae* are closer to *F. sporotrichioides* than those of *F. poae* and *F. kyushuense*. Unfortunately, both *F. poae* and powdery *F. poae* genotypes seemed to have the same geographic and host distribution. This stresses the need for a method which can be used for their distinction.

Fusarium acuminatum sensu lato has recently been split into two subspecies, on the basis of morphological and ecological characteristics, namely *F. acuminatum* subsp. *acuminatum*, and *F. acuminatum* subsp. *armeniacum* (Burgess et al., 1997). Isolates of *F. acuminatum* subsp. *armeniacum* were able to produce larger amounts of T2 and NEO than those produced by *F. acuminatum* spp. *acuminatum* (Logrieco et al., 1992). In addition, some preliminary phylogenetic evidence, apart from the peculiar T2 synthesis capability, suggested a closer genetic affinity between *F. acuminatum* subsp. *armeniacum* and *F. sporotrichioides* than with *F. acuminatum* subsp. *acuminatum*.

Three species were differentiated within *F. sambucinum* on the basis of morphological and toxicological characteristics, namely *F. sambucinum sensu stricto* a producer of DAS, NEO, T2 and ENB; *F. torulosum* a producer of ENB; and *F. venenatum* mainly a DAS producer (Nirenberg, 1995b; Altomare et al., 1995).

F. avenaceum (Fr.) Sacc. is not generally accepted as a homogeneous species. On the basis of the host range and of morphological and ecological characteristics observed in limited Australian environments, this species was recently split into three subspecies, that is *F. avenaceum* subsp. *avenaceum* Sangalang et al.; *F. avenaceum* subsp. *aywerte* Sangalang and Burgess; and *F. avenaceum* subsp. *nurragi* Sommerell and Burgess (Burgess et al., 1994). The restricted niche and the host specialization of the latter two subspecies led to the assumption that only the former could be worldwide in distribution, and all the past references in the literature which related to *F. avenaceum* (Fr.) Sacc. should probably refer to *F. avenaceum* subsp. *avenaceum*. In addition, a number of north European strains of *F. avenaceum* from wheat and barley were not clearly distinguished by cladistic analysis, and even though clustered into several sub-groups, showed no pathogenic specialization on wheat and barley (Yli-Mattila et al., 1997).

Furthermore, notwithstanding the different hosts and ecological conditions that distinguish most European areas, there is only scant evidence of the occurrence of molecular diversity within the population of *F. avenaceum* occurring on small-grain cereals. This situation certainly requires future investigation, especially in relation to strains of *F. tricinctum* which appear to belong to the same geno-species, and which show the same mycotoxin profile, notably in relation to the production of MON (Schütt et al., 1998).

Fusarium nivale is a well known pathogen of cereals, very frequently found among the major fungi included in the species complex causing foot rot and head blight (scab) of small-grain cereals. *F. nivale* is no longer considered to be a *Fusarium* species. It was first placed in the genus *Gerlachia* (*G. nivalis*), then transferred to *Microdochium* as *M. nivale* (teleomorph *Monographella nivalis*) (Gams, 1989). *M. nivale* has a very low, if any, mycotoxin producing ability (Logrieco et al., 1991), and proved incapable of producing the typical *Fusarium* trichothecenes and zearalenone *in vitro* (Chełkowski et al., 1991; Nakajima and Naito, 1995). No typical *Fusarium* toxins were found in naturally *M. nivale*-infected ears of wheat and rye, even with a blight incidence of up to 46% and 65% respectively (Chełkowski et al., 1991).

Fusarium species and mycotoxins associated with fusarium head blight in Europe

Fusarium species occurring on cereal ears in the field can produce many mycotoxins, some of which are of notable importance. This section focuses on the compounds most commonly found in analytical surveys of field or freshly harvested infected kernel samples. The naturally occurring *Fusarium* mycotoxins belong to three main structural groups, that is trichothecenes, ZEN and MON. In addition, beauvericin (BEA) has also been found in naturally infected small-grain cereal and is considered as an emerging toxicological problem. The principal toxicological features of these mycotoxins are briefly outlined by Bottalico (1998).

Fusarium *species associated with FHB in Europe*

A profile of the *Fusarium* complex species associated with FHB in wheat in Europe is presented in Table 1. Regarding the three predominant species, it appears that *F. graminearum* is the most common in moist-warm continental climates, such as central and south-eastern Europe. In contrast, *F. culmorum* and *F. avenaceum* are more often found in maritime and cooler European areas. However, the influence of several factors favourable to the disease can lead to a change in *Fusarium* profile from region to region, as well as from year to year, as exemplified by the following regional situations.

The data on the incidence of FHB on wheat in Italy in the year 2000 (Pasquini et al., 2001), showed an incidence of the disease in the northern regions (Lombardia, and Emilia-Romagna) (average of 4% on both soft and hard wheat), which was lower than that in central regions (Umbria, Toscana, Lazio and Abruzzo) (average of 35% and 25% on soft and hard wheat, respectively). The disease was absent in the southern regions (Molise, Puglia and Sicily). The higher FHB incidence in central regions can be explained by the more pronounced influence of maritime warmer weather. In addition, the data obtained in the year 2000 showed a slightly higher incidence of FHB on soft wheats than on hard wheat. Other data indicate that hard cultivars were more susceptible than soft cultivars (Balmas et al.1999). FHB is also commonly observed on barley cultivated from central (Lazio) to northern (Emilia-Romagna and Lombardia) Italian regions (Delogu et al., 2001).

The species predominantly found associated with FHB of both soft and hard wheat in Italy are *F. graminearum* (mostly Group 2), *F. culmorum* and *F. avenaceum*. A prevalence of *F. graminearum* and *F. culmorum* was observed when weather conditions around the time of flowering were wetter and warmer than normal, while drier weather appears more conducive to *F. culmorum* and less to *F. graminearum*. In particular, *F. graminearum* (32%), *F. avenaceum* (31%), *F. culmorum* (25%) and, to a lesser extent, *F. cerealis*, *F. poae* and *F. tricinctum* were mostly isolated from FHB of soft wheat (Balmas and Corazza, 1994; Casulli et al., 1995). During three years of observations (1994–1996) of FHB in hard wheat throughout Italy, Pancaldi and Torricelli (1999) found *F. culmorum* and *F. graminearum*, associated with *F. avenaceum*, predominant in all three years, and also *F. verticillioides*, *F. tricinctum* and *F. cerealis* in 1994. In 1996, *F. poae* was isolated with high frequency. Balmas et al. (1999) confirmed the predominance of *F. graminearum* and *F. culmorum* from blighted heads of durum and bread wheat and stressed the prevalence of the former in northern Italy and of the latter in central and southern Italy. Moreover, among the other occurring species involved in FHB of hard wheat were reported

F. avenaceum, *F. acuminatum*, *F. cerealis*, *F. chlamydosporum*, *F. semitectum*, *F. compactum*, *F. equiseti*, *F. poae*, *F. solani*, *F. oxysporum*, *F. proliferatum*; *F. subglutinans* and *F. sporotrichioides* (Balmas et al., 1999; Pasquini et al., 2001).

A similar *Fusarium* profile observed in Italy was reported for FHB in wheat in other similar southern European localities, including Portugal, Spain, and France (Maurin and Chenet, 1993; Assemat et al., 1995).

A slightly shifted spectrum of the predominant species, characterized by a partial displacing of *F. graminearum* by *F. avenaceum* and *F. culmorum*, was reported for southern Germany (Bavaria), north-eastern regions of Austria, Croatia, Swiss midlands, and Czech and Slovak Republics. In regions from central to northern Germany an emerging prevalence of *F. poae* and *F. sporotrichioides* was noted, in addition to the three predominant species, together with the occasional appearance of *F. tricinctum*, *F. equiseti*, *F. cerealis*, *F. acuminatum* and *F. solani* (Mauler-Machnik and Suty, 1997; Müller and Reiman, 1997; Schütze et al., 1997). In FHB of durum wheat in Austria, *F. graminearum* (26.8%) was dominant, with *F. avenaceum* (11.1%) *F. poae* (2.6%), *F. culmorum* (1.9%), *F. equiseti* (1.8%), *F. cerealis* (<1%), *F. oxysporum* (<1%) *F. subglutinans* (<1%) and *F. solani* (<1%) (Adler et al., 1995). In the eastern part of Croatia, the following results were found: *F. graminearum* (51.08%), *F. verticillioides* (20.38%), *F. avenaceum* (12.61%), *F. subglutinans* (9.05%), *F. poae* (1.91%) and *F. solani* (1.53%); whereas the species isolated from barley included *F. graminearum* (71.37%), *F. verticillioides* (13.65%), *F. avenaceum* (6.61%), *F. subglutinans* (6.61%), and *F. poae* (1.76%). In the Swiss midlands, the *Fusarium* species most represented in FHB in winter wheat were *F. graminearum* (up to 42% of diseased ears) and *F. avenaceum* (Schachermayr and Fried, 2000). A profile of the *Fusarium* species complex similar to that found in central Europe was reported for the Czech and Slovak Republics where *F. verticillioides* and *F. proliferatum* were isolated from wheat and barley (Srobárová and Pavlová, 1997; Srobárová, 1997). In the Czech Republic the spectrum of species on spring barley in 1997–98 included: *F. culmorum* (70%), *F. poae* (20%), *F. avenaceum* (2%), and to a much lesser extent, *F. stilboides*, *F. aquaeductuum*, *F. merismoides* and *F. gigas*. An increase in FHB severity in spring barley in recent years has been reported, and this change appears to be related to a drastic shift in environmental conditions as well as to agro-technical factors and host genotype. In particular, a different *Fusarium* spp. complex was observed on barley depending on the previous crop: *F. tricinctum* and *F. poae* were dominant after sugar beet, *F. culmorum* and *F. oxysporum* after maize, and *F. culmorum*, *F. poae* and *F. tricinctum* after small-grain cereals.

In cooler maritime areas of the north-western European countries, including north-western areas of France, the Netherlands, Belgium, England, and Scotland, the most common *Fusarium* species involved in FHB in small-grain cereals were *F. culmorum*, *F. graminearum*, *F. avenaceum* and *F. poae*. However, the latest available surveys report an increasing prevalence of *F. culmorum* and a greater importance of *F. poae* and *F. avenaceum*, especially in years less conducive to *F. graminearum* infection (Daamen et al., 1991; Parry et al., 1995; De Nijis et al., 1996). In the United Kingdom, which was characterised by very high FHB in wheat, surveys carried out during 1998 indicated the south west and Wales as the most affected regions, with a prevalence of *F. culmorum* and *F. graminearum* as *Fusarium* causal agents (Polley and Turner, 1995). In the Netherlands, the number of freshly harvested cereal samples infected with *Fusarium* differed considerably between 1991 (34%) and 1993 (83%), possibly due to the humid and dry weather conditions which occurred at flowering. The most predominant species in 1991 were *F. culmorum* and *F. avenaceum* isolated only in northern areas, while in 1993 the most predominant species were *F. poae*, *F. culmorum* and *F. cerealis*, which were isolated throughout the country (De Nijs et al., 1996).

In north-eastern Europe, for example in Poland, the wheat surveys carried out during the late 1990s and covering various climatic areas, indicated the predominance of *F. poae* (64%), followed by *F. tricinctum* (15%), *F. avenaceum* (8%), *F. culmorum* (6%), and *F. graminearum* (4%), together with an emerging occurrence of *F. sporotrichioides*, and a sporadic presence of *F. equiseti* and *F. oxysporum* (Perkowski et al., 1997a; Goliński et al., 1997). However, previous surveys carried out during the 1980s indicated an annual variation in relation to the location and the environmental conditions, showing a wider occurrence of *F. graminearum*, *F. avenaceum* and *F. cerealis* especially on wheat and oats in southern areas, and a greater relevance of *F. poae*, *F. culmorum* and *F. avenaceum* on wheat, triticale and barley in central to northern locations (Perkowski et al., 1990; 1997a). In the seasons between 1985 and 1989, the most frequently isolated species was *F. avenaceum*, both from wheat and triticale, (Lew et al., 1993). Similar results were obtained for freshly harvested kernels of spring barley in the Lublin region, where in three seasons (1986–88) the most frequently isolated *Fusarium* species was *F. avenaceum*, followed by *F. culmorum*, and to a much lesser extent *F. graminearum* (Lacikowa and Kiecana, 1991). Finally, *F. avenaceum* proved to be the dominant *Fusarium* species on triticale. In fact, in 1986 the most important species infecting heads of winter triticale were *F. avenaceum* (39%), *F. culmorum* (20%), *F. graminearum* (14%), and others (6%), including *F. cerealis*, *F. poae* and *F. equisieti*. The following year, 1987, was characterized by a long snowy winter, very conducive to *M. nivale* (64%), and in these conditions the spread of *Fusarium* species was suppressed, with a lower presence of *F. avenaceum* (24%), *F. culmorum* (6%), and *F. graminearum* (5%) (Perkowski et al., 1988). But, in the next period from 1987 to 1989, the percentage of FHB triticale heads infected by *F. avenaceum* was again higher with incidences of 23%, 83%, and 40% (Lew et al., 1993).

In Russia, FHB of small-grain cereals is very widespread and the loss of the wheat crop has reached 25–50%. In north-western regions of Russia, the most frequently isolated species were *F. culmorum*, *F. avenaceum*, *F. tricinctum*, *F. poae*, and *F. sporotrichioides*, while *F. graminearum* was almost absent (Ablova, 1997; Schipilova and Gagkaeva, 1997). In central areas (Moscow region), *F. culmorum*, *F. avenaceum*, *F. sporotrichioides* and *F. poae*, were the dominant species, followed by *F. equiseti*, *F. subglutinans*, *F. tricinctum*, *F. graminearum*, *F. sambucinum*, and *F. acuminatum*. In the warm and humid conditions of the south-eastern European part of Russia (Bielorussia, Krasnodar, Ukraine) and in the Far East, a higher and less variable predominance of *F. graminearum* was reported, together with *F. avenaceum*, *F. poae*, *F. sambucinum* and *F. verticillioides*.

In south-eastern Europe, for example in Hungary, the *Fusarium* species most frequently found in winter wheat samples in the early 1990s were *F. equiseti* (16%), *F. poae* (16%), *F. sporotrichioides* (14%), *F. graminearum* (13%), and *F. semitectum* (10%) and, to a lesser extent, *F. culmorum* (6%), *F. chlamydosporum* (6%), *F. avenaceum* (6%), *F. acuminatum* (5%), and *F. oxysporum* (2%) (Tóth, 1997). However, in earlier 1970s and 1980s surveys, quite different *Fusarium* profiles were recorded, with an alternate predominance of *F. graminearum* and *F. culmorum*, combined with a frequent or rare occurrence of *F. poae*, *F. semitectum* and *F. oxysporum* (Tóth, 1991). In contrast, in the neighbouring northeast areas of Romania and Bulgaria, the most widespread species was *F. culmorum*.

The northern European situation can be exemplified by the following Norwegian surveys, also confirmed by Finnish observations. The *Fusarium* species most frequently encountered during the latest Norwegian surveys of cereal samples collected from the central to the southern part of the country included, in decreasing order, *F. avenaceum*, *F. tricinctum*, *F. poae*, *F. graminearum*, *F. culmorum*, *F. equiseti*, *F. torulosum*, and *F. sporotrichioides* (Langseth et al., 1997). However, a prevalence of *F. tricinctum*, *F. poae* and *F. culmorum* was observed on spring wheat when weather conditions around the time of flowering were drier and warmer than normal. Wetter and colder weather was conducive to *F. avenaceum* and *F. graminearum* and less to *F. culmorum* (Elen et al., 1997). Other less frequently represented species were *F. acuminatum*, *F. cerealis*, *F. flocciferum*, *F. oxysporum*, *F. pallidoroseum*, *F. proliferatum*, *F. sambucinum*, *F. sporotrichioides*, *F. chlamydosporum* and *F. verticillioides* (Langseth et al., 1997; Kosiak et al., 1997). The Danish and Finnish field situations appear similar, as confirmed from the profile of *Fusarium* species reported by Thrane (2000). In particular, the profile of detected *Fusarium* spp. included, in decreasing order, *F. avenaceum* and *F. tricinctum* as the most frequently represented, followed by *F. poae* and *F. culmorum*, and then to a lesser extent, by *F. graminearum*, *F. equiseti*, and *F. sporotrichioides*. However, in Denmark, *F. culmorum* is believed to be the major cause of FHB of barley.

Fusarium *mycotoxins associated with FHB in Europe*

Deoxynivalenol. Field surveys indicate that the mycotoxins most frequently encountered in FHB of wheat throughout the European countries are DON and its derivatives, produced by *F. graminearum* and *F. culmorum*.

A survey carried out by Pascale et al. (2000, 2001) in Italy, on freshly harvested cereal grains in 1998–2000, showed the occurrence of DON in samples of soft wheat from Emilia-Romagna, both in traditionally grown (incidence 5/36, DON average in contaminated samples 0.146 mg kg^{-1}, range 0.055–0.33 mg kg^{-1}), and biologically-grown cultivars (incidence 18/35, DON average 0.070 mg kg^{-1}, range 0.050–0.105 mg kg^{-1}). Levels of DON higher than that in soft wheat were also found in samples of hard wheat collected from the same region in 1998 (incidence 24/26, DON average 0.202 mg kg^{-1}, range 0.060–1.000 mg kg^{-1}).

Deoxynivalenol was also found in samples of *T. dicoccum* grown biologically (incidence 11/13, DON average 0.237 mg kg^{-1}, range 0.125–0.350 mg kg^{-1}), and in samples of *T. spelta* grown biologically (incidence 7/7, DON average 0.119 mg kg^{-1}, range 0.070–0.220 mg kg^{-1}). Moreover, DON was found in untreated samples of triticale (*T. secalotricum*) (incidence 8/10, DON average 0.116 mg kg^{-1}, range 0.050–0.200 mg kg^{-1}), and barley (incidence 2/20, DON average 0.805 mg kg^{-1}, range 0.070–1.540 mg kg^{-1}). A survey carried out in 1999 showed decreasing contamination from northern Italian regions (Lombardia, Emilia Romagna), to central regions (Toscana, Umbria), with a very low DON incidence in southern regions (Puglia, Basilicata, Sicilia). In that year, contamination was, in general, higher, than that of 1998, and the difference in susceptibility of hard wheat to soft wheat was more marked. The data obtained from surveys carried out in 2000, confirmed both the decreasing DON contamination from northern to southern Italian regions, and the greater susceptibility of hard wheat than soft wheat. In particular, the data obtained in Lombardy showed a degree of contamination for hard wheat (incidence 2/2, DON average 3.385 mg kg^{-1}, range 1.610–6.465 mg kg^{-1}) significantly higher than that of soft wheat (incidence 2/2, DON average 1.371 mg kg^{-1}, range 0.380–3.010 mg kg^{-1}).

In Poland, samples of wheat ears naturally infected with *F. graminearum* and *F. culmorum* were contaminated with DON (100%) and 3AcDON (80%) at very high concentrations (up to 30.4 and 29.54 mg kg^{-1}, respectively), and the levels of DON and 3AcDON in the chaff were 5 and 50 times higher than those in the kernels (Visconti et al., 1986). In addition, analysis of cereal grain field samples (200 samples) carried out in 1994–95 in central Poland showed the occurrence of DON, 3-AcDON, and 15-AcDON, in wheat, barley, and oats, with the highest DON level in barley (average of 0.12 mg kg^{-1}) (Chełkowski, 1989b; Perkowski et al., 1990, 1997b; Goliński et al., 1997). A similar analysis of freshly harvested small-grain cereal from western Poland, showed the occurrence of DON at an average concentration of 0.6 and 1.530 mg kg^{-1} in 1996 and 1997. The DON content in kernels was well correlated to scab symptoms, because the species involved in FHB mostly included the mycotoxin-producing species *F. graminearum* (up to 23%) and *F. culmorum* (up to 25%). Finally, samples of triticale kernels from ears infected by *F. graminearum* and *F. culmorum* were all found contaminated by DON (1.6–16.4 mg kg^{-1}) and 3-AcDON (7–1.4 mg kg^{-1}), with higher concentrations in chaff (9.9–33.2 and 5.2–16 mg kg^{-1}, respectively) (Perkowski et al., 1988).

In Austria, 48 freshly harvested kernel samples of durum wheat predominantly infected by *F. graminearum* and with a lower presence of *F. culmorum*, contained high amounts of DON (up to 8.2 mg kg^{-1}), together with low levels of ZEN (up to 0.330 mg kg^{-1}) (Adler et al., 1995). DON was associated with FHB in neighbouring Slovakia and in the Swiss midlands, together with a high occurrence of ZEN (Srobárová and Pavlová, 1997) in ears of winter wheat heavily contaminated with *F. graminearum*.

In Bavaria, Germany, DON was found in 87% of wheat samples at very high levels (mean 3.96 mg kg^{-1}, max 43.8 mg kg^{-1}), together with ZEN, NIV and T2 (Lepschy-von Gleissenthal et al., 1989). In addition, DON levels between 0.27 and 0.36 mg kg^{-1} were found in field samples of barley, rye and oats (Lepschy-von Gleissenthal et al., 1989). In the same country, Obst et al. (1997) observed that the severity of wheat FHB and DON content in scabby grains were correlated with the nature of the preceding crops. Higher DON concentrations were found in wheat following maize for silage (up to 0.3 mg kg^{-1}) and maize for grain (up to 0.5 mg kg^{-1}). Furthermore, it was observed that the occurrence of DON in harvested wheat and rye was higher in conventional farming systems than in organic systems. In the south-western areas of Germany (Baden-Wuerttemberg), DON (up to 4.76 mg kg^{-1}, mean 0.4 mg kg^{-1}) was found in 89% of several freshly harvested barley seed infected by *Fusarium*, together with low levels of 3-AcDON and ZEN in 48 and 68% of the samples, respectively (Müller and Schwadorf, 1993a). In the same south-western area of Germany, a very high contamination of DON (incidence 96%, mean 1.632 mg kg^{-1}, max 20.538 mg kg^{-1}) associated with 3-AcDON (incidence 59%, mean 0.007 mg kg^{-1}, max 0.018 mg kg^{-1}), ZEN, α-ZOH, NIV, T2, and HT2, was found in 84 farm samples of wheat for feed use collected in a year characterised by heavy rainfall in the summer months (Müller and Schwadorf, 1993b).

In Norway, the concentrations of DON in a large number of *Fusarium* infected ear samples of wheat, barley and oats, collected in 1995 under weather conditions that were very conducive to *F. graminearum*, were 2.36, 0.36 and 0.69 mg kg^{-1}, respectively (Elen et al., 1997). Moreover, analyses of 5000 freshly harvested wheat, barley and oats samples surveyed during the period 1988–1996, showed a mean year concentration of DON from 0.035 mg kg^{-1} in 1996 to 1072 mg kg^{-1} in 1988, with a maximum DON content from a single sample of 62.050 mg kg^{-1}. The percentages of oats, barley and wheat containing more than 1 mg kg^{-1} of DON were 12.5, 6.6 and 3.1 respectively. Oats generally contained greater amount of DON in the northern areas in Norway (Trondelag region) than in southern areas, whereas barley and wheat were more affected in southern areas (Langseth and Elen, 1996; 1997).

In other Northern countries, the occurrence of DON was reported in Finland (up to 1.18 mg kg^{-1}) (Hietaniemi and Kumpulainen, 1991), and in the Netherlands (up to 0.5 mg kg^{-1}) together with ZEN (De Nijis et al., 1996). The occurrence of mycotoxins in UK small-grain cereals was generally related to the severity of FHB infection, and highest DON concentrations were found in grains infected with *F. culmorum* and *F. graminearum* (Turner and Jennings, 1997). However, the amount of DON produced was lower than that expected from the severity of ear blight, especially when the predominance of *M. nivale* supressed the infections by toxigenic *Fusarium* species, as reported for winter wheat ear blight surveyed in 1998 in southwest England and Wales.

In southern localities of Russia (Krasnodarski krai), FHB in wheat is widespread (losses up to 25–50%), and high amounts of DON and its derivatives were frequently found in freshly harvested scabby grains. In particular, a very high occurrence of DON (incidence of 100%, range 7.25–36.25 mg kg^{-1} was found in 15 selected kernel samples from several fields of wheat affected by FHB predominantly caused by *F. graminearum*, associated with consistent levels of 4,7-dideoxy-NIV (0.16–1.25 mg kg^{-1}) (Leonov et al., 1990).

Severe infestations of wheat ears by *F. graminearum* are also reported in Bulgaria (scab incidence of 37.2%), with an average DON content of 1.8 mg kg^{-1} in 67% of 140 samples collected after harvest throughout the country (DON average in positive samples 0.180 mg kg^{-1}). Together with DON, 3-AcDON and 15-AcDON were found in 2.1% and 0.7% of the samples, respectively (both up to 0.100 mg kg^{-1}) (Vrabcheva et al., 1996; Vrabcheva and Vrabchev, 1997).

In the Czech Republic, DON and 15-AcDON were found in 71% and 12% of the samples. In one sample, the DON concentration was higher than the limit stipulated in Czech food safety regulations (2 mg kg^{-1}). In addition, levels of DON, sometimes higher than the hygiene limit, were found in spring barley grown after maize, and predominantly contaminated by *F. culmorum*.

Although mould growth was reported in Finnish rye in the field, only small amounts of DON were detected in rye kernels. In autumn 1998, in spite of the wet summer, DON was observed in 70% of rye samples, in the concentration range 0.005–0.050 mg kg^{-1}.

Nivalenol and fusarenone-X (FUS). The occurrence of NIV and FUS, mainly co-occurring with DON, was frequently found in ears of cereals affected by FHB from southern to northern European localities, where they were attributed to the presence of *F. graminearum* NIV-chemotypes and to *F. culmorum* NIV-chemotypes. In addition, NIV and FUS formation has also been attributed to the activity of *F. poae*, especially in Sweden and other northern countries and to *F. cerealis*, particularly in grains from central to northeast Europe (Bottalico et al., 1990; Eriksen and Alexander, 1998).

In Polish samples collected in 1987, NIV (up to 0.01 mg kg^{-1}) and 4,7-dideoxy-NIV (=7-deoxy-DON) (up to 0.15 mg kg^{-1}), were found together with DON, 3-AcDON, 15-AcDON, and ZEN (Perkowski et al., 1990). These findings, which for the first time reported the occurrence of 15-AcDON and 3-AcDON in European wheat, were confirmed during the latest surveys of scabby field grains, particularly those infected with *F. poae*, which showed highest NIV (0.056 mg kg^{-1}) and FUS (0.052 mg kg^{-1}) levels in oats (Goliński et al., 1997; Perkowski et al., 1997a). Moreover, samples of freshly harvested oat kernels collected at different locations in Poland, were contaminated by NIV (0.05 mg kg^{-1} (Chełkowski et al., 2000).

In Germany, in 1987, NIV (0.04–0.29 mg kg^{-1}) was found in 30% of wheat samples (Lepschy-von Gleissenthal et al., 1989). In the southwest area of Germany (Baden-Wuerttemberg), samples of wheat for feed use collected in a year characterised by heavy rainfall in the summer months, showed traces of NIV (incidence 26%, mean 0.009 mg kg^{-1}, max 0.032 mg kg^{-1}) associated with DON, 3-AcDON, ZEN, αZOL, T2, and HT2 (Müller and Schwadorf, 1993b).

A survey of 15 selected field kernel samples of wheat from Krasnodarski Krai in the USSR, predominantly colonized by *F. graminearum*, showed high levels of 4,7-dideoxy-NIV (0.16–1.25 mg kg^{-1}) associated with DON (Leonov et al., 1990).

In UK grains, highest levels of NIV were associated with *F. poae* and *F. culmorum*, while higher DON concentrations were found in grains infected by *F. graminearum* and *F. culmorum* (Turner and Jennings, 1997).

In the Czech Republic, during a screening of 50 wheat samples collected at harvest in 1999, NIV and FUS were found in 69% and 2% of the samples. Concentrations of NIV lower than the legislated limit (2 mg kg^{-1}), were found in spring barley, particularly when cultivated after small-grain cereals predominantly infected by *F. poae*, *F. culmorum* and *F. tricinctum*.

Zearalenone. Zearalenone, mainly produced by *F. graminearum* and *F. culmorum* and commonly associated with DON and its derivatives, is among the mycotoxins most frequently encountered in FHB in small-grain cereals throughout the European countries.

Austrian pre-harvest hard wheat ears, predominantly infected by *F. graminearum* and with a lower presence of *F. culmorum*, contained low levels of ZEN (0.33 mg kg^{-1}) which were associated with high amounts of DON (up to 8.2 mg kg^{-1}) (Adler et al., 1995). In Slovakia, a high occurrence of ZEN, together with DON, was found associated with FHB in wheat mainly infected by *F. graminearum* (Srobárová and Pavlová, 1997).

In Bavaria, Germany, ZEN (up to 1.56 mg kg^{-1}) was found in 58% of wheat samples, mainly associated with DON (Lepschy-von Gleissenthal et al., 1989). It was also observed that the amount of ZEN in harvested wheat was higher in conventional farming systems (up to 0.74 mg kg^{-1}) than in organic systems (up to 0.47 mg kg^{-1}). A survey of wheat collected on farms from the southwest area of Germany, in a year characterised by heavy rain in the summer months, showed high contamination of ZEN (incidence 80%, mean 0.178 mg kg^{-1}, max 8.036 mg kg^{-1}), co-occurring with α-ZOH (incidence 5%, mean 0.023 mg kg^{-1}, max 0.071 mg kg^{-1}), DON, 3-AcDON, NIV, T2, and HT2 (Müller and Schwadorf, 1993b).

In the Netherlands, ZEN (up to 0.2 mg kg^{-1}) was found in freshly harvested cereal samples, together with DON (De Nijis et al., 1996), and in Bulgaria ZEN was found in 69% of 140 wheat samples for human consumption in concentrations up to 0.120 mg kg^{-1} (mean in positive sample of 0.017 mg kg^{-1}) (Vrabcheva et al., 1996).

T-2 toxin derivatives. Epidemics of *F. sporotrichioides* and *F. poae* in cold European localities may lead to the occurrence of T-2 derivatives (T2, HT2, T2ol), and DAS and MAS, respectively. In Poland scabby field grains contained T2 and HT2 in wheat, barley and oats, with the highest concentrations in oats (average 0.302 mg kg^{-1}, predominantly infected by *F. sporotrichioides* (Perkowski et al., 1997a). In addition, T2 (up to 2.4 mg kg^{-1}), HT2 (up to 0.37 mg kg^{-1}) and T2-tetraol (up to 0.21 mg kg^{-1}) were found in barley heads infected by *F. sporotrichiodes* collected in southeast areas of Poland (Perkowski et al., 1997b). Moreover, samples of freshly harvested oats kernels collected at different locations in Poland were contaminated by DAS (0.02 mg kg^{-1}), T2 (0.06 mg kg^{-1}) and HT2 (0.02 mg kg^{-1}) (Chełkowski et al., 2000).

In Bavaria, Germany, cereal samples containing low levels of T2 (0.005–0.60 mg kg^{-1}) were found in 38% of wheat samples (Lepschy-von Gleissenthal et al., 1989). Wheat collected from farms in Baden-Wuerttemberg showed the occurrence of T2 (incidence 26%, mean 0.82 mg kg^{-1}, max 0.249 mg kg^{-1}) associated with HT2 (incidence 7%, mean 0.010 mg kg^{-1}, max 0.020 mg kg^{-1}), DON, 3-AcDON, ZEN, α-ZOH and NIV (Müller and Schwadorf, 1993b).

In the Czech Republic, HT2, T2ol and DAS were found in 28%, 10% and 6% of the wheat samples analysed in 1999. In addition, T2 (0.055 mg kg^{-1}) was found in 1 of 140 samples of freshly harvested wheat from Bulgaria (Vrabcheva et al., 1996).

Moniliformin. Severe infections of *F. avenaceum*, *F. tricinctum*, and to a lesser extent *F. subglutinans*, from central to north-east European countries, were usually responsible for MON occurring in scabby grains (Kostechi et al., 1995). In particular, in Poland, significant levels of MON were found in scabby kernels obtained in 1988 from ears of wheat and triticale (7.2–25.2 and 2.4–5.1 mg kg^{-1}) highly infected by *F. avenaceum* (Lew et al., 1993). MON was also reported in freshly harvested durum wheat in Austria (up to 0.88 mg kg^{-1}) (Adler et al., 1995). In all these surveys, the MON content in kernels was well correlated with the presence of *F. avenaceum* (up to 38%) (Kostechi et al., 1995; Chełkowski et al., 2000).

Beauvericin. Beauvericin was detected as a natural contaminant of wheat kernels collected in Poland (Kostecki et al., 1997) and co-occurred with ENS in Finland (up to 3.5 mg kg^{-1}). In addition, most Finnish strains of *F. avenaceum* and *F. poae* isolated from contaminated wheat samples produced BEA and other esadepsipeptides in high amounts on autoclaved maize (up to 3,703 mg kg^{-1}) (Logrieco et al., 2002).

Fumonisins. It is interesting to note that the occurrence of *F. verticillioides* and *F. proliferatum* in infected wheat ears in Norway (Kosiak et al., 1997), Bulgaria, the Czech and Slovak Republics (Parry et al., 1995; Srobárová and Pavlová, 1997; Srobárová, 1997) Croatia and Hungary (Tóth, 1997), South Russia and Italy (Pasquini et al., 2001, Pancaldi and Alberti, 2001) as well as of toxigenic strains of (*F. verticillioides* and *F. proliferatum*) in freshly harvested wheat and oats kernels from southern European countries (Italy, France, Portugal, Greece, and Turkey) (Bottalico et al., 1989; Infantino et al., 2001), is indicative of the spread of these fumonisin-producing species. As a consequence, it seems that the occurrence of FB_1, until now limited to maize, can be expected in infected wheat and other small grains. The occurrence of strains of *F. proliferatum* able to produce fumonisins B_1, B_2 and B_3 in rye grains (Fadl-Allah et al., 1997) confirms the above indication of the spread of this phytopathogenic species, and of the potential occurrence of its carcinogenic toxins in small grains.

Acknowledgments

This work was supported by COST-835 "Agriculturally Important Toxigenic Fungi" and by EU project (QLK1-CT-1999-01380).

References

Ablova IB (1997) Resistance to head blight in various types of wheat and triticale in Krasnodar environment. Cereal Research Communications 25: 715–720

Adler A, Lew H, Brodacz W, Edinger W and Oberforster M (1995) Occurrence of moniliformin, deoxynivalenol, and zearalenone in durum wheat (*Triticum durum* Desf.). Mycotoxin Research 11: 9–15

Altomare C, Logrieco A, Bottalico A, Mulé G, Moretti A and Evidente A (1995) Production of type A trichothecenes and enniatin B by *Fusarium sambucinum* Fuckel *sensu lato*. Mycopathologia 129: 177–181

Aoki T and O'Donnell K (1999a) Morphological and molecular characterization of *Fusarium pseudograminearum* sp. nov., formerly recognized as Group 1 of *F. graminearum* and PCR primers for its identification. Mycologia 91: 597–609

Aoki T and O'Donnell K (1999b) Morphological characterization of *Gibberella coronicola* sp. nov., obtained through mating experiments of *Fusarium pseudograminearum*. Mycoscience 40: 443–453

Assemat P, Dawson M, Quenin H and Tregou MP (1995) Fusarioses des épis de blé en France. Fréquence et localisation des espèces. Phytoma 479: 22–24

Balmas V and Corazza L (1994) *Fusarium graminearum* Gruppo 2 agente causale della scabbia del frumento duro nell'Italia centrale. L'Informatore Agrario 11: 59–62

Balmas V, Maccaroni M, Santori A and Corazza L (1999) Fusariosi della spiga di cereali autunno-vernini: Indagine epidemiologica e valutazione della resistenza varietale di frumento duro nell'Italia centrale. L'Informatore Agrario 12: 37–42

Bottalico A (1997) Toxigenic *Fusarium* species and their mycotoxins in pre-harvest cereals in Europe. Bulletin of the Institute for Comprehensive Agricultural Sciences Kinki University, Nara, Japan 5: 47–62

Bottalico A (1998) *Fusarium* diseases of cereals: Species complex and related mycotoxins profiles, in Europe. Journal of Plant Pathology 80: 85–103

Bottalico A, Logrieco A and Visconti A (1989) *Fusarium* species and their mycotoxins in infected cereals in the field and in stored grains. In: Chelkowki J (ed) *Fusarium* – Mycotoxins, Taxonomy and Pathogenicity (pp 85–120) Elsevier, Amsterdam, The Netherlands

Bottalico A, Logrieco A and Visconti A (1990) Mycotoxins produced by *Fusarium crookwellense* Burgess, Nelson and Toussoun. Phytopathologia Mediterranea 29: 124–127

Burgess LW, Summerell BA, Bullock S, Gott KP and Backhouse D (1994) Laboratory manual for *Fusarium* research. Department of Crop Science and Royal Botanic Garden, University of Sydney, Australia, 133 pp

Burgess LW, Summerell BA, Backhouse D, Benyon F and Levic J (1997) Biodiversity and population studies in *Fusarium*. Sydowia, special issue 48: 1–11

Casulli F, Pancaldi D, De Lillo E and Alberti I (1995) Observations on wheat crown rot and head blight caused by *Fusarium* spp. in Italy. In: Abstracts of International Seminar on *Fusarium* – Mycotoxins, Taxonomy and Pathogenicity (pp 139–140) May 9–13, 1995, Martina Franca, Italy

Chełkowski J (1989a) *Fusarium* – Mycotoxins, Taxonomy and Pathogenicity. Elsevier, Amsterdam, The Netherlands, 492 pp

Chełkowski J (1989b) Formation of mycotoxins produced by *Fusarium* in heads of wheat, triticale and rye. In: Chelkowki J (ed) *Fusarium* – Mycotoxins, Taxonomy and Pathogenicity (pp 63–84) Elsevier, Amsterdam, The Netherlands

Chełkowski J (1991a) Cereal grains Mycotoxins, Fungi and Quality in Drying and Storage. Elsevier, Amsterdam, The Netherlands

Chełkowski J (1991b) Second European Seminar Fusarium: Mycotoxins, taxonomy, pathogenicity. September 5–7 1990, Poznan, Poland. Mycotoxin Research 7: Part I: 1–96; Part II: 97–187

Chełkowski J, Goliński P, Perkowski J, Visconti A, Rakowska M and W. Wawulinski (1991) Toxinogenicity of *Microdochium nivale* (*Fusarium nivale*) isolates from cereals in Poland. Mycotoxin Research 7: 140–145

Daamen RA, Langerak CJ and Stol W (1991) Surveys of cereal diseases and pests in The Netherlands. III. *Monographella nivalis* and *Fusarium* spp. in winter wheat fields and seed lots. Netherlands Journal of Plant Pathology 97: 105–114

Delogu G, Faccini N, Lavarelli M, Caruso S, Notario T, Piccioni I, Paoletta G, Riccardi M and Signor M (2001) Le principali malattie dell'orzo. L'Informatore Agrario 1: 40–42

D'Mello JPF, MacDonald AMC, Postel D and Hunter EA (1997) 3-Acetyl deoxynivalenol production in a strain of *Fusarium culmorum* insensitive to the fungicide difenoconazole. Mycotoxin Research 13: 73–80

De Nijis M, Soentoro P, Delfgou-Van Asch E, Kamphuis H, Rombouts FM and Notermans SHW (1996) Fungal infection and presence of deoxynivalenol and zearalenone in cereals grown in The Netherlands. Journal of Food Protection 59: 772–777

Elen O, Langseth W, Liu W, Haug G, Skinnes H, Gullord M and Sundheim L (1997) The content of deoxynivalenol and occurrence of *Fusarium* spp. in cereals from field trials in Norway. Cereal Research Communications 25: 585–586

Eriksen GS and Alexander G (1998) *Fusarium* toxins in cereals – a risk assessment. TemaNord Food 1998: 502, Nordic Council of Ministers. Expressen Tryk & Kopicenter, Copenaghen, Denmark, 146 pp

Fadl-Allah E, Stack M, Goth R and Bean G (1997) Production of fumonisins B_1, B_2 and B_3 by *Fusarium proliferatum* isolated from rye grains. Mycotoxin Research 13: 43–48

Gams W (1989) Taxonomy and nomenclature of *Microdochium nivale* (*Fusarium nivale*). In: Chełkowski J (ed) *Fusarium*: Mycotoxins, Taxonomy and Pathogenicity (pp 195–198) Elsevier, Amsterdam, The Netherlands

Gang G, Miedaner T, Schuhmacher U, Schollenberger M and Geiger HH (1998) Deoxynivalenol and nivalenol production by *Fusarium culmorum* isolates differing in aggressiveness toward winter rye. Phytopathology 88: 879–884

Goliński P, Perkowski J, Kostecki M, Grabarkiewicz-Szczesna J and Chełkowski J (1997) *Fusarium* species and *Fusarium* toxins in wheat in Poland – a comparison with neighbour countries. Sydowia (special issue March 1997) 48: 12–22

Hietaniemi V and Kumpulainen J (1991) Contents of *Fusarium* toxins in Finnish and imported grains and feeds. Food Additives and Contaminants 8: 171–182

IARC (1993) IARC Monographs on the Evaluation of Carcinogenic risks to Humans. Vol. 56. Some Naturally Occurring Substances: Food Items and Constituents, Heterocyclic Aromatic Amines and Mycotoxins. International Agency for Research on Cancer, Lyon, France

Infantino A, Conca G, Di Giambattista G, Pucci N and Porta-Puglia A (2001) Lo stato sanitario delle cariossidi di frumento. L'Informatore Agrario 18: 90–91

Kosiak B, Torp M and Thrane U (1997) The occurrence of *Fusarium* spp. in Norwegian grain – A survey. Cereal Research Communications 25: 595–596

Kostechi M, Szczesna J, Chełkowski J and Wisniewska H (1995) Beauvericin and moniliformin production by Polish isolates of *Fusarium subglutinans* and natural co-occurrence of both mycotoxins in cereal grain samples. Microbiologie, Aliments, Nutrition 13: 67–70

Kostecki M, Grabarkiewicz-Szczesna J and Goliński P (1997) Simultaneous analysis of beauvericin and moniliformin in fungal cultures and in cereal grain samples. Mycotoxin Research 13: 17–22

Lacicowa B and Kiecana I (1991) Fusariosis of spring barley cultivated in Lublin region. Mycotoxin Research 7: 128–135

Langseth W and Elen O (1996) Differences between barley, oats and wheat in the occurrence of deoxynivalenol and other trichothecenes in Norwegian grain. Journal of Phytopathology 144: 113–118

Langseth W and Elen O (1997) The occurrence of deoxynivalenol in Norwegian cereals – differences between years and districts, 1988–1996. Acta Agriculturae Scandinavica. Section B, Soil and Plant Science 47: 176–184

Langseth W, Bernhoft A, Rundberget T, Brekke T, Kosiak B and Gareis M (1997) Cytotoxicity and mycotoxin production of *Fusarium* strains isolated from Norwegian grain. Cereal Research Communications 25: 419–421

LBP – Bayerische Landesanstalt für Bodenkultur und Pflanzenbau (2000) Risiken durch den Ährenparasiten *Fusarium graminearum* – Ergebnisse eines LBP-Forschungsverbunds. Bodenkultur und Pflanzenbau Vol 3, 107 pp

Leonov AN, Kononenko GP and Soboleva NA (1990) Production of DON-related trichothecenes by *Fusarium graminearum* Schw. from Krasnodarski krai of the USSR. Mycotoxin Research 6: 54–60

Lepschy-von Gleissenthal J, Dietrich R, Märtlbauer E, Schuster M, Süss A and Terplan G (1989) A survey on the occurrence of *Fusarium* mycotoxins in Bavarian cereals from the 1987 harvest. Zeitschrift für Lebensmittel-Untersuchung und-Forschung 188: 521–526

Lew H, Chełkowski J, Wakulinski W and Edinger W (1993) Moniliformin in wheat and triticale grain. Mycotoxin Research 9: 66–71

Logrieco A, Vesonder RF, Peterson SW and Bottalico A (1991) Reexamination of the taxonomic disposition of and deoxynivalenol production by *Fusarium nivale* NRRL 3289. Mycologia 83: 367–370

Logrieco A, Altomare C, Moretti A and Bottalico A (1992) Cultural and toxigenic variability in *Fusarium acuminatum*. Mycological Research 96: 518–523

Logrieco A, Seifert KA, Leslie JF, Petrini O and Petrini LE (1997) Biodiversity of toxigenic *Fusarium* species. Sydowia special issue 1997 Verlag, Horn, Austria, 228 pp

Logrieco A, Rizzo A, Ferracane L and Ritieni A (2002) Occurrence of beauvericin and enniatins in wheat affected by *Fusarium avenaceum* head blight. Applied Environmental Microbiology 68(1): 82–85

Mauler-Machnik A and Suty A (1997) New findings on the epidemiology, importance and control of *Fusarium* ear blight on wheat. Cereal Research Communications 25: 705–709

Maurin N and Chenet I (1993) Espèces de *Fusarium* présentes dans les cultures de céréales en France. Phytoma 453: 20–22

Mesterházy Á (1997a) Proceedings of the Fifth European *Fusarium* Seminar, Szeged, Hungary, 29 August–5 September 1997. Cereal Research Communications 25(3)/1–2: 231–867

Mesterházy Á (1997b) Methodology of resistance testing and breeding against *Fusarium* head blight in wheat and results of the selection. Cereal Research Communications 25: 631–637

Mesterházy Á, Bartók T, Téren J and Korom Á (1991) Resistance level and toxin contamination in wheat and corn after artificial inoculation. Mycotoxin Research 7/1: 64–67

Miller JD (1995) Fungi and mycotoxins in grains: Implications for stored product research. Journal of Stored Products Research 31: 1–16

Miller JD, Greenhalgh R, Wang YZ and Lu M (1991) Trichothecene chemotype of three *Fusarium* species. Mycologia 83: 121–130

Miller JD and Trenholm HL (1994) Mycotoxins in Grains: Compounds other than Aflatoxins. Eagan Press, St. Paul, Minnesota

Müller HM and Schwadorf K (1993a) Natural occurrence of *Fusarium* toxins in barley grown in a south-western area of Germany. Bulletin of Environmental Contamination and Toxicology 51: 532–537

Müller HM and Schwadorf K (1993b) A survey of the natural occurrence of *Fusarium* toxins in wheat grown in a south-western area of Germany. Mycopathologia 121: 115–121

Müller HM and Reiman J (1997) *Fusarium* toxins in wheat harvested during six years in an area of South West Germany. Natural Toxins 5: 24–30

Nakajima T and Naito S (1995) Reassessment of mycotoxin productivity of *Microdochium nivale* in Japan. Annals of the Phytopathological Society of Japan 61: 357–361

Nirenberg HI (1995a) The European *Fusarium sambucinum* project. Mycopathologia 129: 131–190

Nirenberg HI (1995b) Morphological differentiation of *Fusarium sambucinum* sensu stricto, *F. torulosum* (Berk. & Curt.) Nirenberg comb. nov. and *F. venenatum* Nirenberg sp. nov. Mycopathologia 129: 131–141

Obst A, Lepschy-von Gleissenthall J and Beck R (1997) On the etiology of *Fusarium* head blight of wheat in south Germany – preceding crops, weather conditions for inoculum production and the head infection, proneness of the crop to infection and mycotoxin production. Cereal Research Communications 25: 699–703

Pancaldi D and Torricelli R (1999) Presenza di specie di *Fusarium* su cariossidi di frumento duro in Emilia-Romagna nel triennio '94–'96. L'Informatore Agrario 3: 42–46

Pancaldi D and Alberti I (2001) Le principali malattie su foglia e spiga del frumento. L'Informatore Agrario 20: 63–69

Parry DW, Jenkinson P and McLeod L (1995) *Fusarium* ear blight (scab) in small grain cereals – a review. Plant Pathology 44: 207–238

Pascale M, De Girolamo A, Visconti A and Pancaldi D (2000) Indagine sulla presenza di deossinivalenolo in cereali prodotti in alcune aree del Nord Italia nel 1998. L'Informatore Agrario 10: 68–73

Pascale M, Pancaldi D, Visconti A, Perrone G and Bottalico A (2001) *Fusarium* ear blight, deoxynivalenol and toxigenic *Fusarium* species in selected wheat cultivars assayed all over Italy, in 2000. In: Proceedings of 11th Congress of the Mediterranean Phytopathological Union, 17–20 September 2001 (pp 123–125) University of Evora, Portugal

Pasquini M, Cardone AM, Casini F, Casulli F, Chierico M, Contoli S, Converso R, Corazza L, Del Vecchio A, Iori A, Lombardi T, Lombardo GL, Mazzocchetti A, Natalizi G, Notario T, Pancaldi D, Raimondo I, Ratti C, Riccardi M, Rubies-Autonell C, Sereni L, Vallega V, Virzì N and Vitale S (2001) Monitoraggio delle malattie sui cereali autunno-vernini: le principali malattie del frumento. L'Informatore Agrario 1: 33–39

Perkowski J, Chełkowski J and Wakuliñski W (1988) Deoxynivalenol and 3-acetyldeoxynivalenol and *Fusarium* species in winter triticale. Mycotoxin Research 4: 97–100

Perkowski J, Plattner RD, Goliński P, Vesonder R.F and Chelkowki J (1990) Natural occurrence of Deoxynivalenol, 3-Acetyl-Deoxynivalenol, 15-Acetyl-Deoxynivalenol, Nivalenol, 4,7-Dideoxynivalenol, and Zearalenone in Polish wheat. Mycotoxin Research 6: 7–12

Perkowski J, Stachowiak J, Kiecana I, Goliński P and Chełkowski J (1997a) Natural occurrence of *Fusarium* mycotoxins in Polish cereals. Cereal Research Communications 25: 379–380

Perkowski J, Jelén H, Kiecana I and Goliński P (1997b) Natural contamination of spring barley with group A trichothecene mycotoxins in south-eastern Poland. Food Additives and Contaminants 14: 321–325

Polley RW and Turner JA (1995) Survey of stem base diseases and *Fusarium* ear diseases in winter wheat in England, Wales and Scotland, 1989–1990. Annals of Applied Biology 126: 45–59

Schachermayr G and Fried PM (2000) Fusarioses et mycotoxines: le tour du probléme. Revue Suisse d'Agriculture 32: 241–246

Schipilova NP and Gagkaeva TY (1997) The forms of manifestation of *Fusarium* head blight on the seed and heads of cereal crops. Cereal Research Communications 25: 815–816

Schütt F, Nirenberg HI and Deml G (1998) Moniliformin production in the genus *Fusarium*. Mycotoxin Research 14: 35–40

Schütze A, Oerke EC and Dehne HW (1997) Isolation and differentiation of *Fusarium* spp. and *Microdochium nivale* on winter wheat in Western Germany. Cereal Research Communications 25: 615–616

Srobárová A (1997) Occurence of *Fusarium* spp. from Slovakia maize kernels. Cereal Research Communications 25: 617–618

Srobárová A and Pavlová A (1997) The pathogen localization and ZEN concentration in wheat infected by *Fusarium graminearum* in relation to nutrition. Cereal Research Communications 25: 449–450

Sydenham EW, Marasas WFO, Thiel PG, Shephard GS and Nieuwenhuis JJ (1991) Production of mycotoxins by selected *Fusarium graminearum* and *F. crookwellense* isolates. Food Additives and Contaminants 8: 31–41

Szécsi A and Bartok T (1995) Trichothecene chemotypes of *Fusarium graminearum* from corn in Hungary. Mycotoxin Research 11: 85–92

Thrane U (2000) Mycotoxin producing *Fusarium* species occurring in Danish cereals. DJF Rapport, Markbrug 24: 165–169

Tóth A (1997) Dominance condition of *Fusarium* species occurring in winter wheat kernels in Pest county. Cereal Research Communications 25: 625–627

Turner JE and Jennings P (1997) The effect of increasing humidity on *Fusarium* ear blight and grain quality. Cereal Research Communications 25: 825–826

Visconti A (2001) Problems associated with *Fusarium* mycotoxins in cereals. Bulletin of the Institute for Comprehensive Agricultural Sciences Kinki University, Nara, Japan 9: 39–55

Visconti A, Chełkowski J and Bottalico A (1986) Deoxynivalenol and 3-acetyldeoxynivalenol – mycotoxins associated with wheat head fusariosis in Poland. Mycotoxin Research 2: 59–64

Vrabcheva T, Gessler R, Usleber E and Märtlbauer E (1996) First survey on the natural occurrence of *Fusarium* mycotoxins in Bulgarian wheat. Mycopathologia 136: 47–52

Vrabcheva T and Vrabchev N (1997) *Fusarium* and deoxynivalenol contamination of wheat grown in various regions of Bulgaria. Cereal Research Communications 25: 395–396

Wang Y and Miller JD (1994) Toxin producing potential of *Fusarium graminearum* from China. Acta Mycologica Sinica 13(3): 229–234

Yli-Mattila T, Mironenko NV, Alekhina IA, Hannukkala A and Bulat SA (1997) Universally primed polymerase chain reaction analysis of *Fusarium avenaceum* isolated from wheat and barley in Finland. Agricultural and Food Science in Finland 6: 25–36

Yoshizawa T (1997) Geographic differences in trichothecene occurrence in Japanese wheat and barley. Bulletin of the Institute for Comprehensive Agricultural Sciences Kinki University, Nara, Japan 5: 23–29

European Journal of Plant Pathology **108:** 625–630, 2002.

Deoxynivalenol, nivalenol and moniliformin in wheat samples with head blight (scab) symptoms in Poland (1998–2000)

M. Tomczak[1], H. Wiśniewska[2], Ł. Stępień[2], M. Kostecki[1], J. Chełkowski[2] and P. Goliński[1,2]
[1]*Department of Chemistry, August Cieszkowski Agricultural University, ul. Wojska Polskiego 75, 60-625 Poznan, Poland (Fax: +48618487824; E-mail: ramag@owl.au.poznan.pl);*
[2]*Polish Academy of Sciences, Institute of Plant Genetics, Poznan, Poland*

Key words: Fusarium, head blight, wheat, deoxynivalenol, moniliformin

Abstract

Fusarium head blight (scab) epidemics of wheat occurred in Żuławy (Northern Poland) during 1998 and in Wielkopolska (West) and in Southern regions of Poland in 1999. Four species were identified in wheat heads with scab symptoms: *Fusarium culmorum*, *Fusarium graminearum*, *Fusarium avenaceum* and *Microdochium nivale*. A significant increase in the frequency of *F. graminearum* (between 23% and 38%), was observed, compared to about 10% during the previous decade. The mycotoxins deoxynivalenol (DON), nivalenol (NIV) and moniliformin (MON) in amounts up to 24.3, 14.2 and 1.72 mg kg^{-1}respectively, were identified in kernels samples.

Introduction

Fusarium head blight (scab) of small grain cereals, particularly wheat and barley, was recently found to be a re-emerging disease with significant reduction of yield and contamination of grain with *Fusarium* mycotoxins (Bai and Shaner, 1994; Bottalico, 1998; Chełkowski, 1998; Jones and Mirocha, 1999; McMullen et al., 1997; Parry et al., 1995). Damage from head scab is multifold: reduction in seed quality, contamination with several mycotoxins (mostly deoxynivalenol (DON)), reduced yield, and the formation of discoloured and shrivelled kernels. A salmon-pink to red fungal growth with sporodochia may be seen along the edge of the glumes or at the base of the spikelet.

Scab has been reported during the 20th century and since epidemics on wheat in 1980–1982 in USA and Canada, DON has been recognized as a *Fusarium* mycotoxin contaminating scabby kernels (Chełkowski, 1998; Jones and Mirocha, 1999; McMullen et al., 1997). Later, *Fusarium* head blight caused significant losses during 1991–1995, both in United States and in Canada (McMullen et al., 1997; Jones and Mirocha, 1999).

The last high incidence of small grain cereal scab in Poland was observed in 1974 (Chełkowski, 1998). Since 1985, our research group has evaluated the percentage of heads affected with the disease. Samples of infected heads were collected and examined to identify species of *Fusarium* sporulating in sporodochia on chaff surfaces as well as colonizing kernels. Mycotoxins have been identified and quantified in chaff and kernels (Chełkowski, 1998; Goliński et al., 1996; Perkowski et al., 1991; 1997; Sharman et al., 1991; Visconti et al., 1986; Wakuliński and Chełkowski, 1993).

In previous studies (Chełkowski, 1998; Wakuliński and Chełkowski, 1993) during 1985–1989, the percentage of individual species in affected wheat heads was: *F. culmorum* 30%, *F. avenaceum* 30%, *F. graminearum* 10% and *M. nivale* 30%. In diseased kernels, trichothecene mycotoxins were most frequently present, with DON as the dominating contaminant, followed by its acetylated derivatives and nivalenol (NIV) (Chełkowski, 1998; Perkowski et al., 1991;

Perkowski et al., 1997; Visconti et al., 1986). Occurrence of NIV was of lower significance than DON. Moniliformin (MON) has been frequently found in wheat grain in Poland, since its first detection in 1990 (Goliński et al., 1996; Sharman et al., 1991). *Fusarium* head blight epidemics during 1998, in Żuławy (North of Poland) encouraged us to examine the fungal species causing the disease and to analyse occurrence of mycotoxins (DON, NIV, MON) in kernels in the following three years.

Materials and methods

Samples

Fields of wheat were inspected in July and August in 1998–2000. A short time before harvest, samples of wheat heads were collected: 61 in 1998, 68 in 1999 and 45 in 2000. They were placed in paper bags and transported to the laboratory. In heads containing pink sporodochia, *Fusarium* species were identified using light microscopy according to the manuals of Nelson et al., 1983 and Kwaśna et al., 1991. Heads were threshed using a laboratory threshing machine. DON and NIV were analysed in samples of kernels from heads infected with *F. culmorum* and/or *F. graminearum*, while MON concentrations were analysed in heads infected with *F. avenaceum*.

Chemical analyses

Finely-ground kernels were extracted with acetonitrile–methanol–water (16 : 3 : 1 v/v/v) using 5 ml of the solvent per 1 g of samples. Extracts were de-fated with n-hexane (3 × 50 ml), and were then concentrated. Half of the extract was used for MON analysis while the remaining part was used to analyse DON and NIV. The extract for MON was purified on Florisil columns (Kostecki et al., 1995). Moniliformin content was estimated on Merck 5554 silica-gel thin-layer chromatography plates with developing solvent 2-propanol–butanol–water–ammonium hydroxide (12 : 4 : 1 : 1). After spraying with 3-methyl-2-benzothiazolinonehydrochloride (MBTH), the intensity of the dark spots on the chromatogram was compared with that of the metabolite standard. To analyse the trichothecenes (DON, NIV), extracts were purified by filtration on a column (Celite 545–charcoal Darco G-60–neutral alumina 3 : 9 : 5 w/w/w) conditioned with acetonitrile-water (82 : 18 v/v). DON and NIV were eluted with the same solvent. Moniliformin, DON and NIV were analysed by HPLC using Waters-501 apparatus with C-18 Nova Pak column (3.9 × 300 mm) and a Waters-486 UV detector (λ_{max} = 224 nm for DON and NIV; λ_{max} = 229 nm for MON). DON and NIV were eluted with 25% water solution of methanol (flow rate 0.7 ml/min) after retention times of 11.72 and 7.46, respectively. Detection limits were: DON and NIV = 0.01 mg/kg. Regarding MON, acetonitrile–water (15 : 85; v/v) buffered with 10 ml of 0.1 M K_2HPO_4 in 40% t-butyl-ammonium hydroxide per 1 litre of solvent was applied as mobile phase (flow rate 0.6 ml/min). Retention time of MON was 11.5 min, detection limit 0.01 mg/kg. Positive results (on the basis of retention times) were confirmed by HPLC analysis with internal standards. Recoveries for NIV, DON and MON were: 75, 87 and 90% respectively.

Sequence characterised amplified regions (SCAR) method of identification of Fusarium *species*

Sporodochia from the chaff surface of one head, after the species had been identified by microscopy, were scratched into an Eppendorff tube. DNA was extracted using a modified CTAB method (Doohan et al., 1998). After grinding the material with a pestle, 500 μl of CTAB buffer with 0.4% 2-mercaptoethanol were added, followed by adding 50 μl of chloroform-octanol (24 : 1/vol.). Samples were incubated for 25 min in a water bath set at 65 °C. Subsequently, after addition of 50 μl of chloroform-octanol, samples were mixed, left at room temperature for 10 min and centrifuged 10 min at 11.000 rpm. The aqueous upper phase was transferred to a new Eppendorff tube. DNA was precipitated with ice-cold ethanol (99.98%) and left in the refrigerator for 2 h. The precipitate was centrifuged at 11.000 rpm for 10 min, supernatant was removed and DNA was washed carefully with cold ethanol (96%) and again centrifuged 5 min at 11.000 rpm. DNA was dried and re-dissolved with 100 μl of TE buffer pH = 8.0. The extracts were stored at −20 °C until used.

Species-specific primers and PCR conditions for all three species were described by Schilling et al., 1996 and Chełkowski et al., 1999. Each sample contained 1 U of Taq DNA polymerase (Finnzymes), 2.5 μl of PCR buffer, 12.5 pmol of forward/reverse primers, 2.5 mM of each dNTP and about 10 ng of fungal DNA. Amplification products were electrophoresed in 1.5% agarose gel (Amersham Biotech) with ethidium bromide at 50 V for about 3 h and visualised under UV light and photographed (Syngen UV visualiser).

Results and discussion

In 1998, *Fusarium* head blight was observed in significant number of fields in the Żuławy region, with about 20% of ears exhibiting visible disease symptoms: pink, abundant sporodochia on chaff, shrivelled and discoloured kernels (*Fusarium* damaged kernels = FDK). The following frequencies of fungi were observed in affected heads: *Fusarium graminearum* 20%, *F. culmorum* 16%, *F. avenaceum* 31% and *Microdochium nivale* 33% (Table 1). The identification of the three *Fusarium* species by microscopy was confirmed by SCAR (Table 1). Furthermore in 9 samples, where *Fusarium* species were not identified by using microscopy (because abundant mycelium contained no macroconidia), the SCAR method was used successfully to identify species in infected heads. In the Wielkopolska region, *Fusarium* head blight incidence in 1998 was below 1% of affected heads on most of the fields. However, the frequency of *Fusarium* species in infected wheat heads was similar to that in the Żuławy region (Table 1).

In 1999, epidemics of *Fusarium* head blight occurred in the Wielkopolska region, 25 years after the previous epidemics observed in 1974 (Chełkowski, 1998), with about 20% of affected heads in most of wheat fields. The frequencies of species infecting heads in this region were the following: *F. graminearum* 38%, *F. culmorum* 7%, *F. avenaceum* 52% and *M. nivale* 3% (Table 1). The incidence of FHB in 1999 at Żuławy was low – about 1% of heads were infected with up to 5% in some fields. The frequency of species infecting ears was the following: *F. graminearum* 26%, *F. culmorum* 10 %, *F. avenaceum* 29% and *M. nivale* 35% (Table 1). The significant increase of the frequency of *F. graminearum* in Wielkopolska (38%) and Żuławy (26%) should be emphasised both. In the previous decade, *F. graminearum* had not been identified in any samples originating from the northern part of the country (Chełkowski, 1998; Perkowski et al., 1991; Visconti et al., 1986; Wakuliński and Chełkowski, 1993). DON, NIV and MON were present in kernels from all regions in both years in amounts (Tables 1–3).

In wheat grain samples in Poland, DON and 3-acetyldeoxynivalenol (3AcDON) were, for the first time, identified in 1985 (Visconti et al., 1986), while zearalenone had been found in 1980 (Chełkowski et al., 1983). The frequency and amounts of zearalenone were found to be low (Perkowski et al., 1990; 1991; 1997). A high correlation was found between the percentage of *Fusarium* damaged kernels (FDK) and the amounts of DON in wheat grain samples in consecutive years 1986–1989 (Perkowski et al., 1991).

Moniliformin was identified for the first time in Polish cereals in 1990 (Sharman et al., 1991) and in each subsequent year it was a frequently occurring metabolite in wheat samples, although at low concentrations – significantly below 1 mg kg^{-1} (Chełkowski, 1998; Goliński et al., 1996). In the present survey, the highest amounts of MON were detected in samples originating from the Żuławy region – up to 0.72 mg kg^{-1} in 1998 and 1.72 mg kg^{-1} in 1999. High levels of MON (above 10 mg kg^{-1}) were found only in samples of *Fusarium* damaged kernels (Sharman et al., 1991). Acetylated derivatives of DON 3- and 15-AcDON, detected in previous survey in wheat samples at very low amounts (Perkowski et al., 1990) were not analysed in samples collected at 1998 and 1999.

Taking into consideration their frequency and concentration, the most important mycotoxins detected in field samples of wheat kernels was DON, occurring in 60–65% of samples in 1998 and in 89–95% of samples in 1999 (Table 1). NIV was detected at high concentrations, up to 14.2 mg kg^{-1} only in 1998, when this metabolite was present in 80–85% of samples. Simultaneous co-occurrence of the three mycotoxins was found in a high percentage of positive samples (33% and 35%), during 1998 and 1999.

The results of this survey confirm the possibility of high contamination of wheat kernels with *Fusarium* mycotoxins under epidemic conditions in Poland. Samples collected from fields not affected with scab contained low amounts of mycotoxins – significantly below 1 mg kg^{-1} (Goliński et al., 1996). High frequency of *M. nivale* infecting heads was associated with low concentration of DON, NIV and MON. In 2000 *Fusarium* scab was not observed in the fields because of very dry weather conditions.

Table 1. Natural contamination of wheat samples with *Fusarium* species and *Fusarium* mycotoxins during seasons 1998 and 1999

Year	Region	Distribution of fungal species (%)				Percentage of positive samples and *Fusarium* mycotoxins concentrations								
		*Fusarium** *graminearum*	*Fusarium** *culmorum*	*Fusarium** *avenaceum*	*Microdochium nivale*	DON			NIV			MON		
						Percentage	Concentration ($mg\,kg^{-1}$)		Percentage	Concentration ($mg\,kg^{-1}$)		Percentage	Concentration ($mg\,kg^{-1}$)	
							Range	Average		Range	Average		Range	Average
1998	Żuławy	20	16	31	33	65	0.01–0.76	0.19	85	0.01–0.63	0.18	66	0.01–0.72	0.12
1998	Wielkp.	23	25	25	26	60	0.02–15.3	3.71	80	0.02–14.2	1.08	20	0.01–0.13	0.08
1999	Żuławy	26	10	29	35	95	0.03–14.01	2.27	57	0.04–0.32	0.15	100	0.01–1.72	0.34
1999	Wielkp.	38	7	52	3	89	0.05–24.29	2.61	30	0.02–0.21	0.08	66	0.02–0.85	0.19

*In all 30 head samples, affected by *Fusarium,* identification by microscopy was confirmed using SCAR.

Table 2. Percentage of samples contaminated with three mycotoxins (DON+NIV+MON), two mycotoxins (DON + NIV or DON + MON or NIV + MON) and one *Fusarium* mycotoxin in wheat kernels samples collected in 1998 and 1999

Year	Toxins detected							
	DON, NIV, MON	DON, NIV	DON, MON	NIV, MON	DON	NIV	MON	ND
1998								
Number of samples	20	17	0	8	2	6	3	5
Percentage (%)	32.8	27.9	0	13	3.3	10	5	8
1999								
Number of samples	24	1	25	1	12	0	2	3
Percentage (%)	35.3	1.5	36.8	1.5	17.6	0	2.9	4.4

Table 3. DON, NIV and MON concentrations in three groups of samples collected in 1998 and 1999

Year	Toxins detected and concentration range								
	<1 mg kg^{-1}			1 mg kg^{-1} $\leq$ $\geq$5 mg kg^{-1}			$\geq$5 mg kg^{-1}		
	DON	NIV	MON	DON	NIV	MON	DON	NIV	MON
1998									
Number of samples	55	60	61	3	0	0	3	1	0
Percentage (%)	90.2	98.4	100	4.9	0	0	4.9	1.6	0
1999									
Number of samples	30	68	66	29	0	2	9	0	0
Percentage (%)	44.1	100	97.1	42.7	0	2.9	13.2	0	0

References

Bai G and Shaner G (1994) Scab of wheat: Pospects for control. Plant Disease 78: 760–766

Bottalico A (1998) *Fusarium* diseases of cereals: Species complex and related mycotoxins profiles in Europe. Journal of Plant Pathology 80: 85–103

Chełkowski J, Goliński P, Mańka M, Trojanowska K, Wiewiórowska M and Szebiotko K (1983) Mycotoxins in cereal grain. Part IX. Zearalenone and *Fusaria* in wheat, barley and corn kernels. Die Nahrung (Food) 27: 526–531

Chełkowski J (1998) Distribution of *Fusarium* species and their mycotoxins in cereals grains. In: Sinha KK and Bathnagar (eds) Mycotoxins in Food and Agricultural Safety (pp 45–64) Marcel Dekker, New York

Chełkowski J, Bateman GL and Mirocha CJ (1999) Identification of toxigenic *Fusarium* species using PCR assays. Journal of Phytopathology 147: 307–311

Doohan FM, Parry DW and Jenkinson P (1998) The use of species-specific PCR-based assays to analyse *Fusarium* ear blight of wheat. Plant Pathology 47: 197–205

Goliński P, Perkowski J, Kostecki M, Grabarkiewicz-Szczęsna J and Chełkowski J (1996) *Fusarium* species and *Fusarium* toxins in wheat in Poland – comparison with neighbour countries. Sydowia 48: 12–22

Jones RK and Mirocha CJ (1999) Quality parameters in small grains from Minnesota affected by Fusarium head blight. Plant Disease 83: 506–511

Kostecki M, Grabarkiewicz-Szczęsna J, Chełkowski J and Wiśniewska H (1995) Beauvericin and moniliformin production by Polish isolates of *Fusarium* subglutinans and natural co-occurrence of both mycotoxins in maize samples. Microbiologie-Aliments-Nutrition 13: 67–70

Kwaśna H, Chełkowski J and Zajkowski P (1991) *Fusarium* (Sierpik) Vol XXII MYCOTA, Institute of Botany, Polish Academy of Sciences, Warszawa-Kraków

McMullen M, Jones R and Gallenberg D (1997) Scab of wheat and barley: A re-emerging disease of devastating impact . Plant Disease 81: 1340–1348

Nelson PE, Toussoun TA and Marasas WFO (1983) *Fusarium* species. An illustrated manual for identification. The Pennsylvania State University Press. University Park and London

Parry DW, Jenkinson P and McLeod (1995) *Fusarium* ear blight (scab) in small grain cereals – a review. Plant Pathology 44: 207–238

Perkowski J, Plattner RD, Goliński P, Vesonder RF and Chełkowski J (1990) Natural occurrence of deoxynivalenol, 3-acetylodeoxynivalenol and zearalenone in polish wheat. Mycotoxin Research 6: 7–12

Perkowski J, Chełkowski J, Błażczak P, Snijders CHA and Wakuliński W (1991) A study of correlations between the amount of deoxynivalenol in grain of wheat and triticale and percentage of *Fusarium* damaged kernels. Mycotoxin Research 7A: 102–114

Perkowski J, Stachowiak J, Kiecana I, Goliński P and Chełkowski J (1997) Natural occurrence of *Fusarium* mycotoxins in Polish cereals. Cereal Research Communication 25: 483–487

Schilling AG, Möller EM and Geiger HH (1996) Polymerase chain reaction-based assays for species-specific detection of *Fusarium culmorum*, *F. graminearum* and *F. avenaceum*. Molecular Plant Pathology 86: 515–522

Sharman M, Gilbert J and Chełkowski J (1991) A survey of the occurrence of the mycotoxin moniliformin in cereal samples from sources worldwide. Food Additives and Contaminants 8: 459–466

Visconti A, Chełkowski J and Bottalico A (1986) Deoxynivalenol and 3 acetyldeoxynivalenol – mycotoxins associated with wheat head fusariosis in Poland. Mycotoxin Research 2: 59–64

Wakuliński W and Chełkowski J (1993) *Fusarium* species causing scab of wheat, rye and triticale in Poznań. Hodowla Roślin, Aklimatyzacja i Nasiennictwo 37(4): 137–142

European Journal of Plant Pathology **108:** 631–637, 2002.

Ochratoxin A in cereals, foodstuffs and human plasma

A. Rizzo[1], M. Eskola[1] and F. Atroshi[2]
[1]*National Veterinary and Food Research Institute, P.O. Box 45, 00581 Helsinki, Finland (Fax: +35893931920; E-mail: aldo.rizzo@eela.fi);* [2]*Department of Clinical Medicine, Pharmacology and Toxicology, Faculty of Veterinary Medicine, P.O. Box 57, FIN-00014 Helsinki University, Finland*

Key words: ochratoxin A, penicillic acid, nephropathy

Abstract

Aspergillus and *Penicillium* fungi contaminate cereals and foodstuffs, and can thus introduce ochratoxin A (OTA) into the food chain. In this work, five new isolates of Aspergillus: *A. albertensis*, *A. auricomus*, *A. wentii*, *A. fumigatus* and *A. versicolor*, were found to produce ochratoxin A. Data on the occurrence and the concentration levels of ochratoxin A in European food of vegetable and animal origin are reported. Furthermore, data on the concentration of ochratoxin A in blood of citizens of Western Europe are compared with those of some areas where Balkan endemic nephropathy is endemic. The results of the studies of Stoev and co-workers are reviewed and the possibility that OTA alone cannot be the cause of the Balkan endemic nephropathy is discussed.

Introduction

Ochratoxin A (OTA) is a mycotoxin produced by several species of *Aspergillus* and *Penicillium* fungi. These fungi are ubiquitous, occurring primarily in temperate climates. Their potential for the contamination of cereals and foodstuffs is widespread. The accumulation of OTA by *Aspergillus* and *Penicillium* is affected by the amount of inoculum, the substrate, water activity, moisture content, temperature, incubation time and by the species of the fungus itself. Ochratoxin A was isolated for the first time by van der Merwe et al. (1965) as a metabolite of *Aspergillus ochraceus.* Ochratoxin A is a dihydromethyl-isocoumarin (ochratoxin α) moiety linked via the 7-carboxy group, by an amidic bond, to a molecule of L-β-phenylalanine (Figure 1). Because of its chemical structure, the best method for the analysis of OTA is HPLC linked to a fluorescence detector. Confirmatory methods are (1) formation of the OTA-methylester by reaction of boron trifluoride in methanol; (2) degradation of OTA by carboxypeptidase with formation of ochratoxin α.

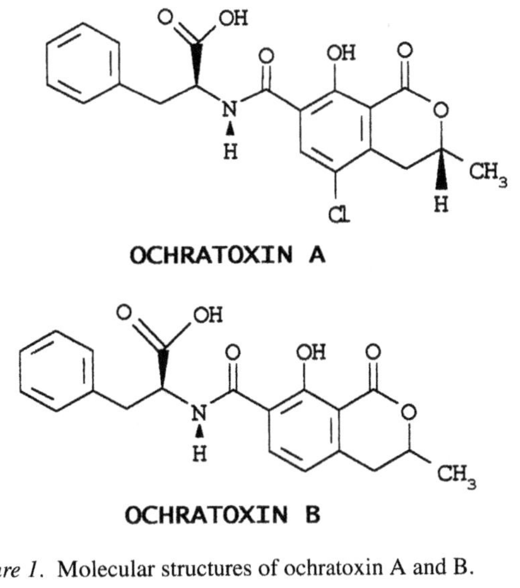

Figure 1. Molecular structures of ochratoxin A and B.

Biological activity of ochratoxin A

Kidneys are probably the target organ for OTA, where it causes nephrotoxic effects, and liver damage in monogastric animals. Mycotoxic porcine nephropathy (MPN) was first recognised in Denmark where it was associated with the use of mouldy rye in feed, and in the Balkan States where human endemic nephropathy is a chronic fatal disease. Aside from nephrotoxicity, OTA has immunosuppressive properties resulting in higher susceptibility to infections, and it is teratogenic and carcinogenic. Furthermore, OTA inhibits protein synthesis by competing with phenylalanine in the phenylalanyl-tRNA synthase-catalysed reaction. The inhibition can be reversed by the presence of phenylalanine (Betina, 1989).

New ochratoxigenic Aspergillus *species*

Recently, Varga et al. (1996) reported three more *Aspergillus* species to be ochratoxigenic. They are: *A. albertensis*, *A. auricomus* and *A. wentii*. Besides producing OTA, all three strains produced ochratoxin B. High-performance thin-layer chromatography was used to analyse cleaned concentrated mycelial extracts. Ochratoxin A was identified under UV light (360 nm) as a bluish-green fluorescent spot with the same R_f value to that of an OTA reference standard. The fluorescence of OTA spots, of both the standard and the extracts, changed to deep blue on treatment of the chromatographic plate with $NaHCO_3$ (5% $NaHCO_3$ in 17% ethanol). The identity of OTA was confirmed by HPLC, by diode array detection and by an immunochemical test (ELISA). *Aspergillus albertensis* was isolated from Canadian cereals, *A. auricomus* from Australian peanuts and *A. wentii* from Indonesian soybeans. One of the three strains of *A. wentii* (IMI 017295), did not produce OTA, even though it had the same origin as that of a toxin-producing strain (ATCC 1023). It is possible that *A. wentii* ATCC 1023 had lost the ability to produce OTA during prolonged maintenance of the culture by periodic transfer on agar slants. Abarca et al. (1997) screened 176 isolates of *Aspergillus*, cultivated on sterilised corn, for their ability to produce OTA. The strains were isolated by the dilution plating technique during the course of a mycological study on animal mixed feeds and raw materials (corn, soybeans and peas). The strains were cultivated on yeast extract-sucrose broth and on sterilised corn. The cultures were extracted with chloroform and analysed by TLC and HPLC. The identity of OTA in samples was confirmed by preparing its methyl ester. One isolate of *A. fumigatus* and one of *A. versicolor* were able to produce this mycotoxin (Abarca et al., 1997). These two species had not previously been reported as ochratoxigenic fungi.

Occurrence of ochratoxin A in cereals and foodstuffs

Ochratoxin A is produced by *Penicillium* species in temperate or cold climates and by a number of *Aspergillus* species in warmer and tropical parts of the world. The best known species of *Aspergillus* that produces OTA in food is *A. alutaceus* (syn. *A. ochraceus*), but OTA is also produced by *A. sulphureus*, *A. sclerotium* or *A. melleus*, which, however, are not very often found in cereals and feeds (Madhyasta et al., 1990; Moss, 1996). *Aspergillus alutaceus* is more common on green coffee beans, cocoa beans, soybeans, peanuts, rice and corn (Kuiper-Goodman and Scott, 1989). *Penicillium verrucosum* is commonly associated with stored cereals and is frequently detected in northern Europe and Canada. Ochratoxin A has been detected in practically all cereals, including corn, barley, wheat, sorghum, rye, oats and rice, and its occurrence is related to climate and especially to harvest and post-harvest conditions. Animal feeds may be contaminated, sometimes at concentrations exceeding 5000 $\mu g\,kg^{-1}$ (Council for Agricultural Science and Technology Cast, 1989). Recently, OTA was detected at high concentrations in dried vine fruits originating from Greece and Turkey (Ministry of the Agriculture, Fisheries and Foods (MAFF), 1997). The concentrations of OTA in European food commodities have been reported by van Egmond and Speijers (1994) and by Pittet (1998). Some of their data are shown in Tables 1–3, while the occurrence of OTA in blood samples of healthy humans is reported in Table 4.

Discussion

The data reported on cereal samples analysed from 1989 to 1991 (Table 1) show rather high concentrations with maximum concentrations reaching the level of milligrams per kilo sample (2400–5410 $\mu g\,kg^{-1}$). The data concerning analyses of OTA in cereals produced during years 1995–1997 (Table 2) show an average incidence of 35%. The highest concentrations are considerably lower than those reported in Table 1. However, the data in Table 2 show that OTA was present in several commodities other than cereals. From the data shown in Table 3, it is also evident that OTA was

Table 1. Ochratoxin A in food commodities of vegetable origin in Europe (from van Egmond and Speijers, 1994)

Commodity	Country	Number of samples analysed	% contamination	Range of concentration (μg kg^{-1})	Reference
Wheat	Yugoslavia[a]	130	8.5	14–135	Pavlovic et al. (1979)
Wheat	Denmark	194	37	0.8–37	Tholstrup and Rasmussen (1990)
Wheat (ecologically grown)	Denmark	36	46	1.2–21	
Wheat	Poland	239	11.7	5–2400	Golinski et al. (1991)
Wheat	Germany	64	1.6	0.4[c]	Frank (1991)
Wheat[b]	Netherlands	38	15	0.1–4.2	van Egmond and Sizoo (1984)
Wheat bran	Germany	84	10.7	6.8[c]	Frank (1991)
Wheat bran	Denmark	57	10.5	5–20	Pedersen and Hansen (1981)
Wheat bran	Denmark	57	68	0.5–12	Tholstrup and Rasmussen (1990)
Wheat bran	Denmark	15	66	0.1–26	
Wheat flour	Poland	137	19.7	3700[c]	Golinski et al. (1991)
Rye	Denmark	267	37	2.5–120	Tholstrup and Rasmussen (1990)
Rye (ecologically grown)	Denmark	53	81	0.7–120	
Rye flour	Poland	78	26.9	5410[c]	Golinski et al. (1991)
Rye	Poland	228	27.2	5–2400	Golinski et al. (1991)
Rye	Germany	64	1.6	0.4[c]	Frank (1991)
Rye[b]	the Netherlands	14	36	0.1–16.8	van Egmond and Sizoo (1984)

[a] Area with endemic nephropathy. [b] Imported grain. [c] Average value.

Table 2. Recent data on the occurrence of ochratoxin A in foods (from Pittet, 1998)

Commodity	Country	Publication year	Number of samples	Incidence (%)	Range of concentration (μg kg^{-1})
Barley	Germany	1997	39	31	0.1–2.7
Wheat	Denmark	1996	520	32	0.05–51
Rye	Denmark	1996	616	42	0.05–121
Oats	Denmark	1996	92	44	0.05–5.6
Wheat	Switzerland	1996	28	54	0.1–10.0
Wheat, barley	UK	1995	1061	4	1–33
Green coffee	Japan	1997	47	30	0.1–17.4
Green coffee	UK	1996	291	38	0.2–27.3
Cocoa powder	UK	1997	20	85	0.2–1.1
Dried vine fruits[a]	UK	1997	60	88	0.2–53.6
Wine	Germany	1996	144	42	0.01–7.0
Spices & herbs[b]	UK	1996	29	86	1.2–50.4

[a] Including raisins, currants and sultanas. [b] Including chilli powder, curry powder, tandoori, ginger, garlic and five spice powders.

present in animal feeds, since 39% of kidney pig samples contained the toxin. The highest concentration of OTA detected was of 240 μg kg^{-1} tissue.

Since OTA is present in several food commodities of vegetable and animal origin, humans and animals can be exposed to OTA through the consumption of contaminated foodstuffs. Ochratoxin A has been detected in human blood and human milk samples and there has been an apparent association between the presence of OTA and Balkan endemic nephropathy. Ochratoxin A has also been associated with urinary tract tumours (Breitholtz-Emanuelsson et al., 1993; Miraglia et al., 1993; Creppy et al., 1993). In 1982, OTA was for the first time detected in blood of inhabitants of regions in Croatia where nephropathy is endemic (Hult et al., 1982). Blood samples, taken

Table 3. Ochratoxin A in food commodities of animal origin in Europe (from van Egmond and Speijers, 1994)

Commodity	Country	Number of samples analysed	% contamination	Range of concentration ($\mu g\,kg^{-1}$)	Reference
Kidney (pig)[a,d]	the Netherlands	46	71	0.2–2.0	van Egmond et al. (1984)
Kidney (pig)[d]	the Netherlands	6	100	0.2–1.0	
Kidney (pig)[a]	the Netherlands	29	7	0.2–0.4	van Egmond et al. (1984)
Kidney (pig)	the Netherlands	6	17	0.2–0.8	
Kidney (pig)[a,e]	the Netherlands	24	71	0.2–240	van Egmond et al. (1984)
Kidney (pig)	Hungary	122	39	2–100	Sandor et al. (1982)
Kidney (pig)	Sweden	129	25	2–104	Rutqvist et al. (1977)
Kidney (pig)	Sweden	90	27	2–88	Josefsson and Moller (1979)
Kidney (pig)	Poland	113	24	1–23	Golinski et al. (1984)
Kidney (pig)	Belgium	95	9.8	0.2 to >80	Rousseau and Van Peteghem (1989)
Kidney (pig)[a]	Czechoslovakia	96	79.2	1–20	Fukal and Marek (1991)
Kidney (pig)	Czechoslovakia	63	1.6	1–5	Fukal and Marek (1991)
Kidney (chicken)[c]	Denmark	—	—	19[b]	Krogh et al. (1976)
Kidney (chicken)[c]	Poland	—	—	1–6	Juskiewicz et al. (1982)
Kidney (quail)[c]	Poland	—	—	80–110	Piskorska-Pliszczinska and Juszikiewicz (1990)
Liver[c]	Germany	—	—	9.1–1.8	Bauer (1988)
Egg (chicken)[c]	Poland	—	—	0.7–13	Juszkiewicz et al. (1982)

[a]From animals with nephropathy. [b]Maximum level detected. [c]Accumulation study. [d]Pigs used for breeding. [e]Originated from Denmark.

from people living in regions where Balkan endemic nephropathy was endemic and from regions where the syndrome was absent, have been investigated for the last ten years. The results of those studies show that OTA was present, not only in blood of people living in regions where nephropathy is endemic, but also in blood of people from regions with no records of nephropathy (Radic et al., 1997). Recently, a new investigation of the occurrence of OTA in human blood of healthy inhabitants living in different regions of Croatia has been carried out by Peraica et al. (1999). The blood samples of the inhabitants of five cities were analysed for the presence of OTA. Positive samples were confirmed by formation of the methyl ester of the toxin. The mean concentration in all samples was $0.39\,ng\,ml^{-1}$ plasma. A concentration of OTA above $0.2\,ng\,ml^{-1}$ (detection limit) was found in 148 of 249 samples (59.4%). In most of the samples, the OTA concentration was between 0.2 and $1.0\,ng\,ml^{-1}$ plasma. The highest value detected was $15.9\,ng\,ml^{-1}$. Considering the data in Table 4, which reports results from investigations carried out on human blood in several European countries, 53% of samples were positive and the average concentration of OTA was $0.5\,ng\,ml^{-1}$. The average value of OTA concentration for the Croatian samples was $0.39\,ng\,ml^{-1}$. According to Peraica et al. (1999), the mean daily intake of OTA in Croatia was $0.53\,ng\,kg^{-1}$ bw/day, which is below the European mean daily intake ($0.9\,ng\,kg^{-1}$ bw/day) (Hoehler, 1998). These results suggest that there is a need to reduce the contamination in cereal-based foodstuffs, and hence decrease the exposure of humans and animals to OTA, not only in areas where Balkan endemic nephropathy occurs, but also in other European countries where the syndrome is absent.

The key question is: if the mean daily intake of OTA in Croatia is lower than that of the Western European countries, then why does the endemic nephropathy exists only in Balkan countries? The answer could be found, perhaps, in the work of Stoev et al. (2001). In that study naturally-occurring porcine nephropathy in Bulgaria was characterised by vascular lesions, renal haemorrhages and enlargement of the renal nodes. These symptoms differ from those of classic MPN, as described in Denmark (Krogh, 1976). Furthermore, the incidence of MPN is higher in Bulgaria than in Denmark by one to two orders of magnitude, even though the overall concentrations of OTA in animal feed (100–200 ppb) and in blood samples from affected animals are substantially lower than those (1–2 ppm) required to reproduce classical MPN of a severity similar to that observed in Bulgaria (Stoev et al., 1998,

Table 4. The occurrence of ochratoxin A in blood samples of healthy humans[a] (from Peraica et al., 1999)

Country	Collection period	Number of positive/analysed (%)	Mean concentration ($ng\ ml^{-1}$)	Range of concentration ($ng\ ml^{-1}$)	Reference
Bulgaria	1984–1990	9/125 (7)	0.2	1.0-10.0	Petkova-Bocharova and Castegnaro (1991)
Canada	1989–1990	45/90 (50)	0.45	0.20–35.33	Frochlich et al. (1991)
	1994	144/144 (100)	0.88	0.29–2.37	Scott et al. (1998)
Czechoslovakia	1990	35/143 (24)	0.14	0.1–1.3	Fukal and Reisnerova (1990)
Czech Republic	1994	734/809 (91)	0.23	0.1–13.7	Malir et al. (1998)
	1995	404/413 (98)	0.24	0.1-1.9	
Croatia	1997	148/249 (59)	0.39	0.2–15.9	Peraica et al. (1999)
Denmark	1986–1988	78/144 (54)	1.8	0.1–13.2	Hald (1991)
France	1991–1992				
Alsace		97/500 (19)		0.1–11.8	Creppy et al. (1993)
Aquitaine		385/2055 (19)		0.1–16.0	
Rhone-Alpe		75/515 (15)		0.1–4.3	
Germany	1977	84/165 (51)	0.79	0.1–14.4	Bauer and Gareis (1987)
	1985	89/141 (68)	0.42	0.1–1.8	Bauer and Gareis (1987)
	1988	142/208 (68)	0.75	0.1–8.0	Hadlock (1993)
Hungary	1995	291/355 (82)		0.2–10.0	Solti et al. (1997)
	1997	213/277 (77)		0.1–1.4	Tapai et al. (1997)
Italy	1992	65/65 (100)	0.53	0.1–2.0	Breitholtz-Emanuelsson et al. (1994)
Japan, Tokyo	1992–1996	156/184 (85)	0.068	0.004–0.278	Ueno et al.1998
Poland	1983–1984	25/397 (6)	0.21	1.0–13.0	Golinski (1987)
	1984–1985	52/668 (8)	0.31	1.0–40.0	
Spain	1996–1998	40/75 (53)	0.71	0.5–4.0	Jimenez et al. (1998)
Switzerland					
North of the Alps		251/252 (100)		0.06–2.14	Zimmerli and Dick (1995)
South of the Alps		116/116 (100)		0.11–6.02	
Sweden	1989				Breitholtz-Emanuelsson et al. (1991)
Visby		29/99 (29)	0.26	0.3–7.0	
Uppsala		3/99 (3)	0.02	0.3–0.8	
Ostersund		6/99 (6)	0.03	0.3–0.8	

[a]Mean concentration is calculated in all samples, and concentration range only in positive samples.

1998a,b). It therefore, appears that Bulgarian MPN cannot be explained by the concentration of OTA alone, but it may be caused by the consumption of multiple mycotoxins. Penicillic acid and other secondary metabolites are produced by fungi of the *A. ochraceus* group, which are the major OTA producers in warm areas, and by *P. verrucosum* which is the main ochratoxigenic fungus in colder areas, such as Scandinavia (Ciegler, 1972; Marquardt and Frolich, 1992). It seems that the penicillic acid has an inhibitory effect on carboxypeptidase activity, an enzyme that is involved in the primary detoxification of OTA in the intestinal tract. In this way penicillic acid could impair the natural detoxification of OTA, thus enhancing its toxicity (Parker et al., 1982). According to the study of Stoev et al. (2001), the synergistic effects between OTA and penicillic acid and possibly other fungal metabolites, may be responsible for the Balkan endemic nephropathy in Bulgaria, which is associated with relatively low levels of OTA in feed ($207.10 \pm 65.14\ \mu g\ kg^{-1}$ and $114.06 \pm 35.79\ \mu g\ kg^{-1}$, Stoev et al., 1998, 1998b). Clearly the interactions between different mycotoxins deserve further study.

References

Abarca ML, Bragulat MR, Castella G, Accensi F and Cabanes FJ (1997) New ochratoxigenic species in the genus *Aspergillus*. Journal of Food Protection 60: 1580–1582

Bauer J and Gareis M (1987) Ochratoxin A in der Nahrungsmittelkette. Zentralblatt Fuer Veterinaermedizin Reihe B. 34(8): 613–627

Bauer J (1988) Disease and depression of productivity in raising swine caused by mycotoxins. [Review] Journal Article. Review, Tutorial, Tierarztliche Praxis – Suppl 3: 40–47
Betina V (1989) Ochratoxins and related dihydroisocoumarins. In: Betina V (ed) Mycotoxins, Chemical, Biological and Environmental Aspects (pp 161–163) Elsevier, Amsterdam–Oxford–New York–Tokyo
Breitholtz-Emanuelsson A, Minervini F, Hult K and Visconti A (1994) Ochratoxin A in human serum samples collected in Southern Italy from healthy individuals and individuals suffering from different kidney disorders. Natural Toxins 2: 366–370
Breitholtz-Emanuelsson A, Olsen M, Dahlbäck A and Hult K (1991) Plasma ochratoxin A levels in three Swedish population surveys using ion-pair HPLC technique. Food Additive and Contaminants 8: 183–192
Breitholtz-Emanuelsson A, Olsen M, Oskarsson A, Palminger I and Hult K (1993) Ochratoxin A in cow's milk and in human milk with corresponding human blood samples. Journal-Association of Official Analytical Chemists 76: 842–846
Ciegler A (1972) Bioproduction of ochratoxin A and penicillic acid by members of the *Aspergillus ochraceus* group. Canadian Journal of Microbiology 18: 631–636
Council for Agricultural Science and Technology Cast (1989) Mycotoxins: Economic and Health Risks. Task Force Report 116, Council for Agricultural Science and Technology, Ames, Iowa, USA, 91 pp
Creppy EE, Castegnaro M, Grosse Y, Meriaux J, Manier C, Montcharmont P and Waller C (1993) Etude de l'ochratoxicose humane dans trois régions de France: Alsace, Aquitaine, et région Rhône-Alpes. In: Creppy EE, Castegnaro M and Dirheimer G (eds) Human Ochratoxicosis and its Pathologies, Vol 321 (pp 147–158) John Libbey, Eurotext, INSERM, Montrouge
Frank HK (1991) Food Contamination by ochratoxin A in Germany. [Meta-Analysis] IARC Scientific Publications (Lyon) 115: 77–81
Frolich AA, Marquardt RR and Ominsky KH (1991) Ochratoxin A as a contaminant in the human food chain: A Canadian perspective. In: Castegnaro M, Plestina R, Dirhaimer G, Chernozemsky and Bartsch H (eds) Mycotoxins, Endemic Nephropathy and Urinary Tract Tumours (pp 139–143) International Agency for Research on Cancer, Lyon
Fukal L and Marek M (1991) A screening method for simultaneous immunochemical detection of aflatoxins and ochratoxin A in cereals and feed. Veterinarni Medicina 36: 251–254
Fukal L and Reisnerova H (1990) Monitoring of aflatoxins and ochratoxin A in Czechoslovak human sera by immunoassay. Bulletin of Environmental Contamination and Toxicology 44: 345–349
Golinski P, Hult K, Grabarkiewicz-Seczesna J, Chelkowski J, Kneblewski P and Szebiotko K (1984) Mycotoxic porcine nephropathy and spontaneous occurrence of ochratoxin A residues in kidneys and blood of polish swine. Applied and Environmental Microbiology 47: 1210–1212
Golinski P (1987) Ochratoxin A in human organism as a result of food and feed contamination. Rocz AR Poznaniu 168: 1–61
Golinski P, Grabarkiewicz-Szczesna J, Chelkowski J, Hult K and Kostecki M (1991) Possible sources of ochratoxin A in human blood in Poland. IARC Scientific Publications (Lyon) 115: 153–158
Hadlok RM (1993) Human ochratoxicosis in Germany updating 1993. In: Creppy EE, Castegnaro M and Dirheimer G (eds) Human Ochratoxicosis and its Pathologies, Vol 321 (pp 141–145) John Libbey, Eurotext, INSERM, Montrouge
Hald B (1991) Ochratoxin A in human blood in European countries. In: Castegnaro M, Plestina R, Dirheimer G, Chernozemsky IN and Bartsch H (eds) Mycotoxins, Endemic Nephropathy, and Urinary Tract Tumours, IARC Scientific Publications (Lyon) 115: 1159–1164
Hoehler D (1998) Ochratoxin A in food and feed: Occurrence, legislation and mode of action. Zeitschrift für Ernährungswissenschaft 37: 2–12
Hult K, Plestina R, Habazin-Novak V, Radic B and Ceovic S (1982) Ochratoxin A in human blood and balkan endemic nephropathy. Archives of Toxicology 51: 313–321
Jimenez AM, Lopez de Cerain A, Gonzales-Peñas E, Bello J, Betbeder AM and Creppy EE (1998) Exposure to ochratoxin A in Europe: Comparison with a region of Northern Spain. Journal of Toxicology-Toxin Reviews 17: 479–491
Josefsson E and Moller T (1979) High pressure liquid chromatographic determination of ochratoxin A and zearalenone in cereals. Journal-Association of Official Analytical Chemists 62: 1165–1168
Juszkiewicz T, Piskorska-Pliszczynska J and Wisnieweska H (1982) Mycotoxins and phycotoxins. In: Proceedings of the 5th International IUPAC Symposium (pp 122–125) Technical University of Vienna Publication, Vienna
Krogh P, Elling F, Hald B, Jylling B, Petersen VE, Skadhauge E and Svendsen CK (1976) Experimental avian nephropathy. Changes of renal function and structure induced by ochratoxin A-contaminated feed. Acta Pathologica et Microbiologica Scandinavica – Section A, Pathology 84: 215–221
Krog P (1976) Mycotoxic nephropathy. In: Krog P (ed) Advances in Veterinary Science and Comparative Medicine. Vol 20 (pp 147–170) Academic Press, New York
Kuiper-Goodman T and Scott PM (1989) Review – risk assessment of the mycotoxin ochratoxin A. Biomedical and Environmental Sciences 2: 179–248
Madhyasta SM, Marquardt RR, Frohlich AA, Platford G and Abramson D (1990) Effect of different cereal and oilseed substrates on the growth and production of toxins by *Aspergillus alutaceus* and *Penicillium verrucosum*. Journal of Agricultural and Food Chemistry 38: 1506–1510
Malir F, Jergeova Z, Severa J, Cerna M, Smid J, Betbeder AM, Baudrimont I and Creppy EE (1998) The level of ochratoxin A in blood serum of adults in the Czech Republic. Review of Medical and Veterinary Mycology 149: 710
Marquardt RR and Frolich AA (1992) A review of recent advances in understanding ochratoxicosis. Journal of Animal Science 70: 3968–3988
Ministry of the Agriculture, Fisheries and Foods (MAFF), November 1997: Survey of aflatoxins and ochratoxin in cereals and retail products. Foods Surveillance Information Sheet No. 130
Miraglia M, Brera C, Corneli S and De Dominicis R (1993) Ochratoxin A in Italy: Status of knowledge and perspectives.

In: Creppy EE, Castegnaro M, Dirheimer G (eds) Human Ochratoxicosis and its Pathologies, Vol 231 (pp 129–139) John Libbey, Eurotext, INSERM, Montrouge

Moss MO (1996) Mode of formation of ochratoxin A. Food Additive and Contaminants 13(Suppl): 5–9

Parker RW, Phillips TD, Kubena LF, Russel LH and Heidelbaugh ND (1982) Inhibition of pancreatic carboxypeptidase A: A possible mechanism of interaction between penicillic acid and ochratoxin A. Journal of Environmental Science and Health B 17: 77–91

Pavlovic M, Plestina R, Krogh P (1979): Ochratoxin A contamination of foodstuffs in an area with Balkan (endemic) nephropathy. Acta Pathologica et Microbiologica Scandinavica – Section B, Microbiology 87: 243–246

Pedersen E and Hansen HN (1981) Stat. Levnedsm. Institute Report F81002. Copenhagen

Peraica M, Domijan AM, Fuchs R, Lucic A and Radic B (1999) The occurrence of ochratoxina A in blood in general population of Croatia. Toxicology Letters 110: 105–112

Petkova-Bocharova T and Castegnaro M (1991) Ochratoxin A in human blood in relation to Balkan endemic nephropathy and urinary tract tumours in Bulgaria. In: Castegnaro M, Plestina R, Dirheimer G, Chernozemsky IN and Bartsch H (eds) Mycotoxins, Endemic Nephropathy and Urinary Tract Tumours, IARC Scientific Publications (Lyon) 115: 135–137

Piskorska-Pliszczynska J and Juszkiewicz T (1990) Tissue deposition and passage into eggs of ochratoxin A in Japanese quail. Journal of Environmental Pathology, Toxicology and Oncology 10: 8–10

Pittet A (1998) Natural occurrence of mycotoxins in food and feeds – an updated review. Revue de Medecine Veterinaire 149: 479–492

Radic B, Fuchs R, Peraica M and Lucic A (1997) Ochratoxin A in human sera in the area with endemic nephropathy in Croatia. Toxicology Letters 91: 105–109

Rousseau DM and Van Peteghem CH (1989) Spontaneous occurrence of ochratoxin A residues in porcine kidneys in Belgium. Bulletin of Environmental Contamination and Toxicology 42: 181–186

Rutqvist L, Bjorklund NE, Hult K and Gatenbeck S (1977) Spontaneous occurrence of ochratoxin residues in kidneys of fattening pigs. Zentralblatt fur Veterinarmedizin – Reihe A 24: 402–408

Sandor G, Glavits R, Vajda L, Vanyi A and Krogh P (1982) Mycotoxins and phycotoxins. In: Proceedings of the 5th International IUPAC Symposium (pp 349–352) Technical University of Vienna Publ., Vienna

Scott PM, Kanhere SR, Lau BPY, Lewis DA, Hayward S, Ryan JJ and Kuiper-Goodman T (1998) Survey of Canadian human blood plasma for ochratoxin A. Food Additives and Contaminants 15: 555–562

Solti L, Salamon F, Barna-Vetro I, Gyongyosi A, Szabo E and Wolfling A (1997) Ochratoxin A content of human seradetermined by a sensitive ELISA. Journal of Analytical Toxicology 21: 44–48

Stoev SD, Vitanov S, Anguelov G, Petkova-Bocharova T and Creppy EE (2001) Experimental mycotoxic nephropathy in pigs provoked by a diet containing ochratoxin A and penicillic acid. Veterinary Research Communications 25: 205–223

Stoev SD, Hald B and Mantle PG (1998) Porcine nephropathy in Bulgaria: A progressive syndrome complex of uncertain (mycotoxin) etiology. The Veterinary Record 142: 190–194

Stoev SD, Grozeva N and Hald B (1998a) Ultrastructural and toxicological investigations in spontaneous cases of porcine nephropathy in Bulgaria. Veterinarski Arhiv 68: 39–49

Stoev SD, Stoeva J, Anguelov G, Hald B, Creppy EE and Radic B (1998b) Haematological, biochemical and toxicological investigations in spontaneous cases with different frequency of porcine nephropathy in Bulgaria. Journal of Veterinary Medicine Series A 45: 229–236

Tapai K, Teren J and Mesterhazy A (1997) Ochratoxin A in the sera of blood donors and ill persons. Cereal Research Communications 25: 307–308

Tholstrup B and Rasmussen G (1990) Publication Levnedsmiddelstyrelsen, Kopenhagen (in Danish)

Ueno Y, Maki S, Lin J, Furuya M, Sugiura Y and Kawamura O (1998) A 4-year study of plasma ochratoxin A in a selected population in Tokyo by immunoassay and immunoaffinity column-linked HPLC. Food & Chemical Toxicology 36: 445–449

van der Merwe KJ, Stein PS and Fourie L (1965) Ochratoxin A, a toxic metabolite produced by *Aspergillus ochraceus* Wilh. Nature (London) 205: 1112–1113

van Egmond HP and Sizoo EA (1984) RIVM Report No. 647705001. National Institute of Public Health and Environmental Protection, Bilthoven, the Netherlands (in Dutch)

van Egmond HP, Van der Molen EJ and Paulsch WE (1984) RIVM Report No. 647704001. National Institute of Public Health and Environmental Protection, Bilthoven, the Netherlands (in Dutch)

van Egmond HP and Speijers GJA (1994) Survey of data on the incidence and levels of ochratoxin A in food and animal feed world-wide. Journal of Natural Toxins 3: 125–144

Varga J, Kevei E, Rinyu E, Teren J and Kozakiewic Z (1996) Ochratoxin production by *Aspergillus* species. Applied and Environmental Microbiology 62: 4461–4464

Zimmerli B and Dick R (1995) Determination of ochratoxin A at the ppt level in human blood, milk and some foodstuffs by high-performance liquid chromatography with enhanced fluorescence detection and immunoaffinity column cleanup: Methodology and Swiss data. Journal of Chromatography B 666: 85–99

European Journal of Plant Pathology **108:** 639–643, 2002.

Ochratoxin A in grapes and wine

Paola Battilani[1] and Amedeo Pietri[2]
[1]*Institute of Entomology and Plant Pathology,* [2]*Institute of Food Science and Nutrition, Faculty of Agriculture, Catholic University, Via E. Parmense 84, 29100 Piacenza, Italy (Fax: +390523599256; E-mail: paola.battilani@pc.unicatt.it)*

Key words: fungi, ochratoxin A, grapes, wine

Abstract

The mycotoxin ochratoxin A is a potent nephrotoxin and a possible human carcinogen. It occurs in a variety of plant products, including wine, grape juice and dried vine fruits. Several surveys have shown that the range of ochratoxin A contents detected in wine produced in Europe varied between 0.01 and 3.4 $\mu g\, l^{-1}$. Both incidence and concentration of the toxin were higher in wines from southern regions and increased in the order white < rosè < red. In Italy, field trials were conducted in 1999 and 2000 to study fungi associated with grapes and their ability to produce ochratoxin. *Aspergillus* and/or *Penicillium* strains were present on grapes, starting from setting in a few vineyards. The highest level of grape colonisation was found at early veraison in 1999 and at ripening in 2000. In both years, 95% of strains belonged to the genus *Aspergillus. Aspergillus niger* aggregate was dominant, with about 50% of the ochratoxin-positive strains identified as *A carbonarius*. Other authors have confirmed the relevance of these fungi and underlined the contribution of *A. carbonarius* to the ochratoxin contamination of wine. This species is very invasive and colonises and penetrates berries, even without skin damage. It emerges that temperature, rain and relative humidity are the main factors that influence ochratoxin production in grapes.

Introduction

The mycotoxin ochratoxin A (OTA) is a potent nephrotoxin to all animal species tested, with the exception of mature ruminants. OTA is carcinogenic to rodents and possesses teratogenic, immunotoxic and possibly neurotoxic and genotoxic properties. Furthermore, it has been implicated as a factor in the human disease Balkan Endemic Nephropathy and the development of urinary tract tumours in humans. In 1993, IARC classified OTA as a possible human carcinogen (group 2B).

The worldwide occurrence of OTA contamination of raw agricultural products has been amply documented. It has been found in a variety of plant products such as cereals, coffee beans, beans and pulses (Kuiper-Goodman and Scott, 1989; Micco et al., 1989; Pohland et al., 1992; Stegen et al., 1997; Jørgensen, 1998). OTA has also been detected in beverages such as beer, wine and grape juice (Majerus and Otteneder, 1996; Zimmerli and Dick, 1996).

Ochratoxin A in grape and grape derivatives

Several surveys in different European countries, in Morocco, Japan and Australia confirmed the frequent presence of OTA on grape products and wine (MAFF, 1997; Ueno, 1998; Burdaspal and Legarda, 1999; MAFF, 1999; Cholmakov-Bodechtel et al., 2000; Festas et al., 2000; Gareis et al., 2000; Majerus et al., 2000; Otteneder and Majerus, 2000; Stockley, 2000; Tateo et al., 2000; Filali et al, 2001; Markaki et al., 2001; Pietri et al., 2001). The range of OTA content in wine produced in Europe varied between 0.01 and 3.4 $\mu g\, l^{-1}$. Values higher than 0.5 $\mu g\, l^{-1}$ were reported by several European authors (Burdaspal and Legarda, 1999; Majerus et al., 2000; Markaki et al., 2001; Pietri et al., 2001), and also from Morocco (Filali et al., 2001). Only Ospital et al., (1998) in France and Festas et al. (2000) in Portugal, reported lower levels or absence of OTA, at least in certain good quality wines. A few samples of dessert wine contained between 1 and

3.9 $\mu g\,l^{-1}$ (Burdaspal and Legarda, 1999; Pietri et al., 2001).

Grape juice, especially from red grapes (Cholmakov-Bodechtel et al., 2000) has been highlighted as an important source of OTA in children's diets. Values reported are between 1.16 and 2.32 $\mu g\,l^{-1}$ (Majerus et al., 2000; Filali et al., 2001). The highest OTA content, among grape and its derivatives, was measured in dried vine fruits (MAFF, 1997) with more than 40 $\mu g\,kg^{-1}$ (MacDonald et al., 1999; MAFF, 1999); OTA was also found in vinegar, especially balsamic vinegar (Markaki et al., 2001), although the level was low (0.2 $\mu g\,l^{-1}$). The high number of reports published during the last 5 years give evidence of the problems related to the presence of OTA in grapes and its processed products for their contribution to the total daily intake, both in adults and children. The interest in this subject is also confirmed by the efforts devoted to develop precise and accurate analytical methods for this toxin (Zimmerli and Dick, 1996a; Tateo et al., 1999; Festas et al., 2000; Visconti et al., 2000; Markaki et al., 2001; Soleas et al., 2001).

All the studies cited are primarily aimed at quantifying OTA, while little information is available about the origin of the toxin. A gradient was observed through regions and wine colour; both OTA incidence and concentration were higher in products from southern regions and increased in the order white < rosè < red (Majerus and Otteneder, 1996; Ospital et al., 1998; Otteneder and Majerus, 2000; Markaki et al., 2001; Pietri et al., 2001). According to Majerus and Otteneder (1996), the latter gradient may be connected with longer mash standing. Zimmerli and Dick (1996) suggest that OTA found in wines of southern European origin is probably formed after the harvest of the grapes, but prior to the alcoholic fermentation. These provisional conclusions were followed by further studies that more carefully considered the fungi responsible for the synthesis of this toxin and their presence in field and post-harvest.

Fungi responsible of the presence of OTA in grape

Fungi responsible of the presence of OTA have been studied especially on cereals, where *Penicillium verrucosum* and *Aspergillus ochraceus* (formerly known as *Aspergillus alutaceus*) are considered the main producers. Until 1998, they were also believed to be responsible for the production of the toxin in grapes (Ospital et al., 1998), but OTA-producing *A. carbonarius* and *A. niger* were identified in dried vine fruits in 1999 (Codex Alimentarius Commission, 1999).

In Italy, field trials were conducted in 1999 and 2000, to study fungi associated with grapes and their ability to produce OTA in two grape-growing areas in the north and in the south of the country. Different farming systems, representative of their own area, were taken into account. Ten bunches were collected from each vineyard at two growth stages in 1999 (early veraison and ripening) and at four growth stages in 2000 (setting, berries increase, early veraison and ripening). They were sub-sampled to isolate *Aspergillus* and *Penicillium* spp. Fungal isolates were purified, characterised and stored on suitable media; afterwards, they were tested for OTA production.

Aspergillus and/or *Penicillium* strains were present on grapes, starting from setting, in a few vineyards. One month later, these fungi were detected in more vineyards, and at early veraison and ripening they were present in all vineyards. The highest level of colonisation was found at early veraison in 1999 and at ripening in 2000 (Battilani et al., 2000, 2001a). These observations are relevant, both because these genera are usually considered post-harvest moulds and because they were all isolated from berries which showed no symptoms. In 1999 and in 2000, 95% of strains belonged to the genus *Aspergillus*. The identification was carried out for some representative strains and, according to these observations, *A. niger* aggregate was dominant, with about 50% of the OTA-positive strains identified as *A. carbonarius*. *Aspergillus fumigatus* and *P. pinophilum* were also identified among the OTA producers (Battilani et al., 2000, 2001a). Other authors have confirmed the relevance of these fungi. In fact, in a study on table grapes carried out in South America, *Aspergillus* section *nigri* moulds were frequently isolated during grape storage and *A. niger* var. *niger* and *A. carbonarius* were identified as potential OTA producers (Da Rocha et al., 2000). In other studies carried out in Spain, Abarca et al. (2001, 2001a) isolated black *Aspergilli* from dried vine fruit and Cabanes et al. (2001) underlined the contribution of *A. carbonarius* to the OTA contamination of wine. In contrast, a specific research programme conducted in Portugal on mycotoxin production by fungi isolated from grapes, pointed out the absence of OTA-producing fungi (Abrunhosa et al., 2001).

OTA content in grapes

The same 10 bunches, collected in each of the nine vineyards, were randomly sampled (Battilani et al., 2000, 2001a) and the berries were analysed for OTA content by HPLC. OTA was found in significant quantities only in 1999 in southern vineyards, starting from veraison: the concentration had increased at harvesting. Substantial differences in OTA content were observed between vineyards, even when the cultivar was the same, and they had been trained in a similar way. Differences in fungicide treatments in the different vineyards were acknowledged, both in number of sprays and in active ingredients used, but the OTA content in grape did not show any clear relationship with these factors. Cultural factors have not shown a clear effect on OTA content in grape, the most remarkable factors being the year and the geographic area (Battilani et al., 2001b). In conclusion, it emerges that temperature (north–south difference), rain and relative humidity (1999–2000 difference) are the main factors that influence OTA production in grapes.

Behaviour of OTA-producing fungi on grape berries

Knowledge of the infection cycle of *Aspergillus* and *Penicillium* is scarce and completely absent as far as grapes are concerned. Preliminary information was obtained by Battilani et al. (2001). Experimental trials were conducted to determine how different OTA-producing *Aspergillus* and *Penicillium* strains colonise grapes. This research consisted of a field and an *in vitro* trial. The former investigated the occurrence of the toxin on the skin and pulp of naturally-infected berries picked in September 1999. The latter tested the ability of strains of *A. carbonarius*, *A. fumigatus* and *P. pinophilum* to colonise, penetrate and develop inside intact and artificially damaged berries. The results showed that even if the skin of grape berries seems to be the more frequently contaminated tissue, OTA was also present in the pulp. The fungal species involved determine the pattern of toxin contamination through berry tissues. In fact, as shown by artificial inoculation, berries were susceptible to OTA-producing fungi early in their growth stage, at least starting from one month after setting, but the amount of fungal growth depended on the species involved. *A. carbonarius* was very invasive; it colonised and penetrated berries, even without skin damage. In contrast, *A. fumigatus* only colonised the pulp in damaged berries and *P. pinophilum* apparently never grew inside berries.

The amount of OTA produced was clearly influenced by the status of the berries and especially that of the fungus. The high content detected in damaged berries inoculated with *A. carbonarius* (30 $\mu g\,kg^{-1}$ produced in 7 days, about 1000 times greater than produced by *A. fumigatus* under the same conditions), is related not only to the intrinsic toxigenic characteristics of the strain, but also to its aggressiveness. In fact, the difference in OTA production by the fungi in liquid medium was not very different (26,910, 20,226 and 16,028 $ng\,kg^{-1}$, respectively, for *A. carbonarius, A. fumigatus* and *P. pinophilum*; Battilani et al., unpublished data).

Conclusions

The need of further research on factors influencing growth and OTA synthesis on grape of OTA-producing fungi is evident. It will allow preventative actions, able to reduce OTA in grape to the minimum possible level to be defined. Up to now, only corrective actions have been studied. Satisfactory results have been obtained using additives able to adsorb OTA, even if with adverse effects on the colour and polyphenol indexes of the wine (Dumeau and Trione, 2000). While waiting for further information, which could be useful to reduce the risk of the presence of OTA and its derivatives in grapes and wine, some encouraging news has come from Punam Jeswal and Jeswal (1998) on antidotal action of grape juice on the negative hepatorenal effects of OTA. They observed that concurrent administration of berry and leaf juice of the common grape to mice, together with OTA, significantly reduced the hepatic and renal damage caused by ingestion of the mycotoxin. In addition, none of these animals showed the formation of hepatocarcinoma, whereas 25% of the animals receiving only OTA developed well-differentiated renal carcinoma and hepatic lesions.

References

Abarca ML, Accensi F, Bragulat MR and Cabanes FJ (2001) Current importance of ochratoxin A-producing *Aspergillus* spp. Journal of Food Protection 64: 903–906

Abarca ML, Accensi F, Bragulat MR and Cabanes FJ (2001a) Ochratoxin A-producing black *Aspergilli* in dried vine fruits. In: Book of Abstracts 'Bioactive Fungal Metabolites – Impact and Exploitation' (p 38) Swansea, Wales

Abrunhosa L, Paterson RRM, Kozakiewicz Z, Lima N and Venancio A (2001) Mycotoxin production from fungi isolated from grapes. Letters in Applied Microbiology 32: 240–242

Battilani P, Giorni P, Bertuzzi T and Pietri A (2001) Preliminary results on ochratoxin A production by fungi invading grape berries. Proceedings 11th Congress of Mediterranean Phytopathological Union and 3rd Congress of Sociedade Portuguesa de Fitopatologia' Evora Portugal, 2001: 120–122

Battilani P, Giorni P, Languasco L, Pietri A and Bertuzzi T (2001a) Dynamics of fungi responsible for ochratoxin A in grape. Results from a survey on their presence in Italy in 2000. In: Book of Abstracts 'Bioactive Fungal Metabolites – Impact and Exploitation' (p 47) Swansea, Wales

Battilani P, Giorni P and Pietri A (2001b) Role of cultural factors on the content of ochratoxin A in grape. Journal of Plant Pathology 83: 231

Battilani P, Languasco L, Pietri A, Bertuzzi T and Giorni P (2000) Ochratoxin A in grape and wine: Causes and conditions of production. Preliminary results. Proceedings of the 5th Congress of the European Foundation for Plant Pathology, Giardini Naxos, Italy, 2000 Abstract 502P, 89. Journal of Plant Pathology, 553–555

Burdaspal PA and Legarda TM (1999) Ochratoxin A in wines and grape products originating from Spain and other European countries. Alimentaria 36: 107–113

Cabanes FJ, Accensi F, Bragulat MR, Abarca ML, Minguez S and Pons A (2001). *Aspergillus carbonarius* is a source of Ochratoxin A-contamination in wine. In: Book of Abstracts 'Bioactive Fungal Metabolites – Impact and Exploitation' (p 51) Swansea, Wales

Cholmakov-Bodechtel C, Wolff J, Gareis M, Bresch H, Engel G, Majerus P, Rosner H and Schneider R (2000) Ochratoxin A: Representative food consumption survey and epidemiological analysis. Archiv fur-Lebensmittelhygiene 51: 111–115

Codex Alimentarius Commission (1999) Codex Committee on Food Additives and Contaminants, 31st session, The Hague, the Netherlands, 22–26 March 1999

Da Rocha CAR, Fraga M, Djalva Da Nobrega Santana M, Combina M, Saenz MA, Magnoli CE and Dalcero MA (2000) Mycological aspects of potential ochratoxin producers in table grapes from two different countries. Proceedings of the X International IUPAC Symposium on Mycotoxins and Phycotoxins, Guaruja, Brasil (Abstract)

Dumeau F and Trione D (2000) Processing and ochratoxin A in wines. Vignevini 27: 79–81

Festas I, Herbert P, Santos L, Cabral M, Barros P and Alves A (2000) Ochratoxin A in some Portuguese wines: Method validation and screening in Port Wine and Vinho Verde. American Journal of Enology and Viticulture 51: 150–154

Filali A, Ouammi L, Betbeder AM, Baudrimont I, Soulaymani R, Benayada A and Creppy EE (2001) Ochratoxin A in beverages from Morocco: A preliminary survey. Food Additives and Contaminants 18: 565–568

Gareis M, Rosner H and Ehrhardt S (2000) Blood serum levels of ochratoxin A and nutrition habits. Archiv fur-Lebensmittelhygiene 51: 108–110

Jørgensen K (1998) Survey of pork, poultry, coffee, beer and pulses for ochratoxin A. Food Additives and Contaminants 15: 550–554

Kuiper-Goodman T and Scott PM (1989) Risk assessment of the mycotoxin ochratoxin A. Biomedical and Environmental Science 2: 179–248

MacDonald S, Wilson P, Barnes K, Damant A, Massey R, Mortby E and Shepherd MJ (1999) Ochratoxin A in dried vine fruit: Method development and survey. Food Additives and Contaminants 16: 253–260

Majerus P and Otteneder H (1996) Detection and occurrence of ochratoxin A in wine and grape juice. Deutusche Lebensmittel-Rundschau 92: 388–390

Majerus P, Bresch H and Otteneder H (2000) Ochratoxin A in wines, fruit juices and seasonings. Archiv fur Lebensmittelhygiene 51: 95–97

Markaki P, Delpont Binet C, Grosso F and Dragacci S (2001) Determination of ochratoxin A in red wine and vinegar by immunoaffinity high-pressure liquid chromatography. Journal of Food Protection 64: 533–537

Micco C, Grossi M, Miraglia M and Brera C (1989) A study of the contamination by ochratoxin A of green and roasted beans. Food Additives and Contaminants 6: 333–339

Ministry of Agriculture, Fisheries and Food – UK (1999) 1998 survey of retail products for ochratoxin A. Food-Surveillance-Information-Sheet No. 185: 1–36

Ministry of Agriculture, Fisheries and Food – UK (1997) Food Survey of aflatoxins and ochratoxin A in cereals and retail products. Surveillance Information Sheet No. 130: 1–30

Ospital M, Cazabeil JM, Betbeder AM, Tricard C, Creppy E and Medina B (1998) L'Ochratoxine A dans les vins. Œnologie 169: 16–19

Otteneder H and Majerus P (2000) Occurrence of ochratoxin A in wines: Influence of the type of wine and its geographical origin. Food Additives and Contaminants 17: 793–798

Pietri A, Bertuzzi T, Pallaroni L and Piva G (2001) Occurrence of ochratoxin A in Italian wines. Food Additives and Contaminants 18: 647–654

Pohland AE, Nesheim S and Friedman L (1992) Ochratoxin A: A review. Pure and Applied Chemistry 64: 1029–1046

Punam Jeswal and Jeswal P (1998) Antidotal effect of grape juice (*Vitis vinifera*) on ochratoxin A caused hepatorenal carcinogenesis in mice (*Mus musculus*) Cytobios 93: 123–128

Soleas GJ, Yan J and Goldberg DM (2001) Assay of ochratoxin A in wine and beer by high-pressure liquid chromatography-photodiode array and gas chromatography-mass selective detection. Journal of Agricultural and Food Chemistry 49: 2733–2740

Stegen G, Jrissen U, Pittet A, Saccon M, Steiner W, Vincenzi M, Winkler M, Zapp J and Schlatter CHR (1997) Screening of European coffee final products for occurrence of ochratoxin AA. Food Additives and Contaminants 14: 211–216

Stockley CS (2000) Ochratoxin A – a metabolite on the agenda for the global wine industry. Australian-Grapegrower & Winemaker No. 438a: 111–112

Tateo F, Bodoni M, Fuso Nerini A, Lubian E, Martello S and Commissati I (1999) Ochratoxin A detection in wines. Industrie delle Bevande 164: 592–596

Ueno Y (1998) Residue and risk of ochratoxin A in human plasma and beverages in Japan. Mycotoxins 47: 25–32

Visconti A, Pascale M and Centonze G (2000) Determination of ochratoxin A in wine by means of immunoaffinity column clean-up and HPLC. Journal of Chromatography A 864: 89–101.

Zimmerli B and Dick R (1996a) Study on the repeated use of commercial immunoaffinity columns. Mitteilungen aus dem Gebiete der Lebensmitteluntersuchung und Hygiene 87: 732–742

Zimmerli B and Dick R (1996) Ochratoxin A in table wine and grape-juice: Occurrence and risk assessment. Food Additives and Contaminants 13: 655–668

European Journal of Plant Pathology **108:** 645–651, 2002.

Ear rot susceptibility and mycotoxin contamination of maize hybrids inoculated with *Fusarium* species under field conditions*

Michelangelo Pascale[1], Angelo Visconti[1] and Jerzy Chelkowski[2]
[1]*Institute of Sciences of Food Production, CNR, Viale L. Einaudi 51, 70125 Bari, Italy (Fax: +390805486063; E-mail: m.pascale@area.ba.cnr.it);* [2]*Institute of Plant Genetics, Polish Academy of Sciences, Strzeszynska 34, 60-479 Poznan, Poland*

Key words: beauvericin, ear rot, fumonisins, fusaproliferin, *Fusarium* infection, maize hybrids, resistance

Abstract

The development of new maize hybrids with resistance to *Fusarium* infection is an effective means of minimizing the risk of mycotoxin contamination. Several maize hybrids have been investigated for *Fusarium* ear rot and accumulation of fumonisin B_1 (FB_1), fumonisin B_2 (FB_2), beauvericin (BEA) and fusaproliferin (FP) after artificial inoculation in the field with toxigenic strains of *Fusarium verticillioides* and *Fusarium proliferatum.* The year of inoculation had a significant influence on the disease severity and mycotoxin accumulation in maize kernels. Of all the hybrids tested, only Mona exhibited resistance to ear rot caused by *F. verticillioides* and produced low levels of fumonisins during three years of experiments. In *Fusarium*-damaged kernels (FDK), fumonisin B_1, fumonisin B_2, beauvericin and fusaproliferin were detected at concentrations much higher (up to 10–20 times) than in healthy-looking kernels (HLK). Animal and human exposure to these mycotoxins can be drastically reduced by removing mouldy and visibly damaged kernels from the commodity.

Introduction

Fusarium verticillioides (synonym *F. moniliforme*) and *F. proliferatum* are important maize pathogens causing stalk and ear rot. They can produce a variety of mycotoxins including fumonisins, beauvericin and fusaproliferin both in pre-harvested and in stored products (Bottalico et al., 1989; Chulze et al., 1998; Logrieco et al., 1995; Moretti et al., 1997; Shephard et al., 1999).

Fumonisins are a group of structurally related mycotoxins occurring in maize and maize-based products intended for animal and human consumption worldwide (Bullerman, 1996; Visconti, 1996). At present, at least fifteen fumonisin analogues have been characterized, but only three of these, fumonisin B_1 (FB_1), fumonisin B_2 (FB_2) and fumonisin B_3 (FB_3), are abundant in naturally contaminated products. FB_1 generally occurs at levels higher than FB_2 or FB_3. Fumonisin B_1 (and to some extent FB_2) causes equine leukoencephalomalacia, porcine pulmonary oedema and liver or kidney cancer in experimental rats or mice. Moreover, fumonisins have been statistically associated with an increased risk of oesophageal cancer in humans consuming contaminated maize in some regions of South Africa (IPCS, 2000; Marasas, 1995; US NPT, 1999).

Beauvericin (BEA) is a mycotoxin produced by several *Fusarium* species, mainly by *F. proliferatum*, *F. semitectum* and *F. subglutinans* (Logrieco et al., 1998; Moretti et al., 1997; Shephard et al., 1999). It has insecticidal activity and is highly toxic towards *Artemia salina* larvae and murine and human cell lines, and induces apoptosis (Grove and Pople, 1980; Krska et al., 1997; Ojcius et al., 1991). It has been found as a natural contaminant of maize in Italy, Austria, Poland, South Africa and USA (Krska et al., 1997;

*Dedicated to Prof Antonio Graniti on the occasion of his 75th birthday.

Logrieco et al., 1993; Munkvold et al., 1998; Ritieni et al., 1997b; Shephard et al., 1999).

Fusaproliferin (FP), a recently characterized metabolite of *F. proliferatum* and *F. subglutinans* (Logrieco et al., 1996; Moretti et al., 1997; Shephard et al., 1999), exhibits toxicity to *Artemia salina*, to insect cells and human B lymphocytes (Logrieco et al., 1996), and was highly teratogenic on chicken embryos (Ritieni et al., 1997a).

The natural co-occurrence of fusaproliferin, beauvericin and fumonisins in pre-harvest maize samples contaminated by *Fusarium* species has been reported in Italy, South Africa and Iowa (Munkvold et al., 1998; Ritieni et al., 1997b; Shephard et al., 1999).

The production of maize hybrids with resistance to *Fusarium* infection and consequent mycotoxin accumulation is an efficient method for decreasing the risk of mycotoxin contamination. The results of studies carried out for several years jointly by the CNR Institute of Toxins and Mycotoxins, Bari (Italy) and the Institute of Plant Genetics, Polish Academy of Sciences, Poznan (Poland) on the susceptibility of different maize hybrids to *F. verticillioides* and *F. proliferatum* infection are summarized in this mini-review in relation to the accumulation of fumonisins, beauvericin and fusaproliferin in maize kernels after artificial infection in the field.

Materials and methods

Twenty-nine maize hybrids were inoculated with toxigenic strains of *F. verticillioides* (KF 1795 and KF 1214) and *F. proliferatum* (ITEM 1752), in one to four seasons, under field conditions. In particular, 14 hybrids were inoculated with *F. verticillioides* during the years 1992–1995 and 15 hybrids with *F. proliferatum* during the years 1996, 1997 and 1999 (Table 1).

Maize hybrids were sown in May, in identical field trials, at Radzikow, 30 km west of Warsaw, Poland. Hybrids were selected for relatively similar maturity dates and different susceptibilities to *Fusarium* ear rot. One-hundred and sixty plants of each hybrid were sown in four rows (5 m long), of which 10 were inoculated with the fungal strains. Fungal inoculation was performed with toothpicks placed in the maize ear according to Hart et al. (1982), at the medium silking stage, one week after silk emergence. Ears from inoculated plants and from control plants (non-inoculated plants) were harvested during the third week of October and transported to the laboratory. Disease assessments were made on 10 inoculated ears and on 10 control ears. A five degree scale was used to evaluate the infection of each ear (Chelkowski et al., 1989). The *Fusarium* ear rot index (*Fi*), ranging from 0 to 500, was calculated using the following equation:

$$Fi = \left(\sum F \times 100\right) / n$$

where F is the disease score of individual ears (1–5) and n is the number of scored ears. Kernels from the same ears were separated in two fractions: *Fusarium*-damaged kernels (FDK, i.e. visibly mouldy kernels) and healthy-looking kernels (HLK, i.e. symptomless kernels). Separate analyses of FDK

Table 1. Maize hybrids tested in different years in *Fusarium* inoculation experiments

Fusarium verticillioides		*Fusarium proliferatum*	
Hybrid	Inoculation year	Hybrid	Inoculation year
KLG 2210	1992, 1993, 1994, 1995	BURAN	1996, 1997, 1999
SMOLIMAG	1992, 1993, 1994, 1995	ELA	1996, 1997, 1999
MONA	1993, 1994, 1995	EWA	1996, 1997, 1999
RUTEN	1992, 1993, 1995	KLG 2210	1996, 1997, 1999
ZENIT	1992, 1993, 1994	MILPA	1996, 1997, 1999
BETULISA	1992, 1995	MONA	1996, 1997, 1999
RAHBE 90102	1992, 1993	JANINA	1996, 1999
ANNA	1994	RAH 895	1996, 1997
BURAN	1994	RUTEN	1996, 1999
ELA	1994	KASIA	1999
MILPA	1994	OLENKA	1999
RAHBE 921573	1995	RAH595	1997
RAHBE 86101	1992	RAH996	1997
SMH 4792	1994	SMH8195	1997
		WILGA	1999

and HLK were performed for fumonisins, beauvericin and fusaproliferin (Pascale et al., 1999). Meteorological data (mean monthly temperatures and monthly precipitations) were obtained from the Meteorological Station located in the Radzikow experimental station.

Results

Results of disease assessments on maize hybrids inoculated with *F. verticillioides* and *F. proliferatum* under field conditions in different years are summarized in Tables 2 and 3.

Inoculation with F. verticillioides. The inoculation of hybrids with *F. verticillioides* resulted in different degrees of ear rot (from 0 to 350) (Table 2). Moreover, the artificial infection of the same hybrids for two or more years gave rise to significant differences in ear rot disease. Hybrids Ruten and Smolimag showed low susceptibility to infection in 1992 and 1993, but they exhibited higher disease scores in 1994 and 1995. Hybrids Ela, Zenit and RAHBE 90102

Table 2. Fusarium ear rot index (disease score) and percentage of *Fusarium*-damaged kernels (FDK) in maize hybrids inoculated with *Fusarium verticillioides* in different years in Poland

Hybrid	*Fusarium* ear rot index (Fi)*				*Fusarium*-damaged kernels (%)			
	1992	1993	1994	1995	1992	1993	1994	1995
SMOLIMAG	60	60	130	200	1	2	5	7
KLG 2210	40	130	110	80	<1	2	2	3
ZENIT	140	230	311	—	2	24	31	—
RUTEN	100	0	—	120	<1	0	—	2
MONA	—	30	110	60	—	2	3	1
RAHBE 90102	240	230	—	—	3	9	—	—
BETULISA	120	—	—	140	2	—	—	9
RAHBE 86101	140	—	—	—	1	—	—	—
RAHBE 921573	—	—	—	170	—	—	—	15
ELA	—	—	350	—	—	—	60	—
MILPA	—	—	220	—	—	—	26	—
ANNA	—	—	133	—	—	—	10	—
SMH 4792	—	—	125	—	—	—	5	—
BURAN	—	—	125	—	—	—	6	—

— not tested. *Fi = 0–500 (0 = healthy cob, 500 = totally rotted cob).

Table 3. Fusarium ear rot index (disease score) and percentage of *Fusarium*-damaged kernels (FDK) in maize hybrids inoculated with *Fusarium proliferatum* in different years in Poland

Hybrid	*Fusarium* ear rot index (Fi)*			*Fusarium*-damaged kernels (%)		
	1996	1997	1999	1996	1997	1999
MONA	170	280	360	11	17	39
BURAN	230	270	410	19	27	37
MILPA	270	250	425	26	17	31
ELA	280	370	450	24	52	54
KLG 2210	340	230	500	37	12	64
EWA	370	360	480	38	37	71
RAH 895	180	280	—	16	28	—
JANINA	210	—	400	25	—	65
RUTEN	240	—	437	24	—	56
SMH 8195	—	260	—	—	18	—
RAH 595	—	270	—	—	20	—
RAH 996	—	310	—	—	30	—
WILGA	—	—	300	—		27
KASIA	—	—	400	—		29
OLENKA	—	—	460	—		40

— not tested. *Fi = 0–500 (0 = healthy cob, 500 = totally rotted cob).

were very susceptible, with high disease scores (up to 350) and a high percentage of FDK (up to 60%). Hybrid Mona was found to be the only one of 14 tested to exhibit low susceptibility in all seasons with about 1.8% FDK in inoculated ears (Pascale et al., 1997).

Inoculation with F. proliferatum. Significant differences in the degree of *Fusarium* ear rot disease were observed in hybrids inoculated with *F. proliferatum* (Table 3). *Fusarium* ear rot index (Fi) ranged from 170 to 370 in 1996, from 250 to 370 in 1997 and from 300 to 500 in 1999; the percentage of *Fusarium*-damaged kernels (FDK) ranged from 11% to 38% in 1996, from 12% to 52% in 1997 and from 27% to 71% in 1999. The hybrids grown in 1999 showed a higher susceptibility to fungal infection compared to the other two years. Moreover, for the same hybrid, different disease scores were observed during two or three years of investigation. In particular, most hybrids (Mona, Buran, Ela, RAH 895, Janina and Ruten) showed higher susceptibility to *Fusarium* ear rot in 1997 and 1999 than in 1996, whereas the hybrids Milpa, Ewa and KLG 2210 were more susceptible to infection in 1996 and 1999 than 1997. In particular, hybrid KLG 2210 showed high disease scores in 1996 and in 1999 (340 and 500, respectively), but exhibited the lowest *Fusarium* ear rot index (230) in 1997 (Pascale et al., 2001).

Mycotoxin accumulation

Results for accumulation of fumonisin B_1 and B_2 in maize kernels after inoculation with *F. verticillioides* are reported in Table 4. The amounts of FB_1, FB_2, beauvericin (BEA) and fusaproliferin (FP) accumulated in maize kernels after inoculation with *F. proliferatum* are reported in Table 5.

Inoculation with F. verticillioides. Fumonisin concentrations in maize kernels ranged from 0.1 to 93.2 $\mu g\ g^{-1}$ for FB_1 and from 0.02 to 15.8 $\mu g\ g^{-1}$ for FB_2. FB_2 was detected at levels lower than FB_1 in all samples tested with the FB_2/FB_1 ratio ranging from 0.13 to 0.43 (0.29 average for all samples). Largest amounts of fumonisins accumulated in kernels of hybrid Zenit, examined for three seasons (up to 37.4 $\mu g\ g^{-1}$ of FB_1 and 8.0 $\mu g\ g^{-1}$ of FB_2), and in hybrid Ela (93.2 $\mu g\ g^{-1}$ of FB_1 and 15.8 $\mu g\ g^{-1}$ of FB_2) inoculated only in 1994. The hybrid Mona contained low levels of fumonisins during three consecutive seasons ranging from 0.1 to 1.7 $\mu g\ g^{-1}$ and from 0.02 to 0.4 $\mu g\ g^{-1}$ of FB_1 and FB_2, respectively. However, significantly higher amounts of fumonisins were detected in grains of hybrids inoculated in the 1994 and 1995 seasons, compared to grains of the same hybrids inoculated in the 1992 and 1993 seasons. Results of the analysis of control ears (non-inoculated) showed that the amount of FB_1 did not exceed 0.04 $\mu g\ g^{-1}$ and FB_2 was not detected.

Table 4. Fumonisin (FB_1 and FB_2) contamination in kernels of maize hybrids inoculated with *Fusarium verticillioides* in different years in Poland

Hybrid	Fumonisin content ($\mu g\ g^{-1}$)*							
	1992		1993		1994		1995	
	FB_1	FB_2	FB_1	FB_2	FB_1	FB_2	FB_1	FB_2
SMOLIMAG	0.2	0.1	0.9	0.2	6.4	1.8	9.6	3.1
KLG 2210	0.4	0.1	0.4	0.1	3.7	1.0	3.4	1.0
ZENIT	2.5	0.8	10.0	2.6	37.4	8.0	—	—
RUTEN	0.9	0.3	0.2	0.02	—	—	5.4	1.6
MONA	—	—	0.1	0.02	1.7	0.4	1.3	0.4
RAHBE 90102	4.3	1.6	6.8	2.4	—	—	—	—
BETULISA	2.2	0.9	—	—	—	—	4.2	1.3
RAHBE 86101	2.6	1.1	—	—	—	—	—	—
RAHBE 921573	—	—	—	—	—	—	14.9	4.6
ELA	—	—	—	—	93.2	15.8	—	—
MILPA	—	—	—	—	12.8	3.7	—	—
ANNA	—	—	—	—	8.1	2.0	—	—
SMH 4792	—	—	—	—	3.9	1.1	—	—
BURAN	—	—	—	—	2.5	0.8	—	—

— not tested. *Weighted averages derived from separate analyses of *Fusarium*-damaged kernels and healthy looking kernels.

Table 5. Fumonisins (FB_1, FB_2), beauvericin (BEA) and fusaproliferin (FP) contamination in maize hybrids inoculated with *Fusarium proliferatum* in different years in Poland

Hybrid	Mycotoxin content ($\mu g\ g^{-1}$)*											
	1996				1997				1999			
	FB_1	FB_2	BEA	FP	FB_1	FB_2	BEA	FP	FB_1	FB_2	BEA	FP
MONA	18.0	1.6	0.8	0.3	66.4	5.3	3.0	1.3	99.6	11.1	9.0	2.0
BURAN	72.4	6.2	1.2	1.8	67.5	6.3	3.9	1.3	122.2	12.3	16.5	1.9
MILPA	54.5	4.3	2.5	0.5	41.1	3.6	4.3	0.4	76.0	7.4	5.5	0.8
ELA	46.3	4.1	1.0	1.1	125.3	11.1	7.4	3.9	173.8	15.9	12.2	4.1
KLG 2210	97.7	8.3	4.2	1.5	26.4	1.9	1.3	0.4	231.9	26.5	18.0	6.4
EWA	130.4	10.7	3.0	3.6	100.1	8.6	7.1	2.8	148.3	15.9	12.1	5.3
RAH 895	20.7	1.6	0.2	0.6	93.1	6.8	2.9	1.4	—	—	—	—
JANINA	38.2	0.4	0.4	0.5	—	—	—	—	210.4	25.3	19.6	4.9
RUTEN	11.4	1.6	1.6	1.1	—	—	—	—	138.7	13.4	15.1	2.8
SMH 8195	—	—	—	—	34.1	2.7	2.3	0.9	—	—	—	—
RAH 595	—	—	—	—	40.8	3.2	2.5	0.8	—	—	—	—
RAH 996	—	—	—	—	50.8	4.2	3.1	1.1	—	—	—	—
WILGA	—	—	—	—	—	—	—	—	29.6	2.8	1.7	0.7
KASIA	—	—	—	—	—	—	—	—	79.5	9.7	8.6	1.5
OLENKA	—	—		—	—	—	—	—	124.2	12.2	14.2	3.0

— not tested. *Weighted averages derived from separate analyses of *Fusarium*-damaged kernels and healthy looking kernels.

The fumonisin content in *Fusarium*-damaged kernels varied greatly, from 5.1 $\mu g\ g^{-1}$ FB_1 and 1.4 $\mu g\ g^{-1}$ FB_2 to 196.0 $\mu g\ g^{-1}$ FB_1 and 62.0 $\mu g\ g^{-1}$ FB_2. Sixteen samples (57%) contained fumonisins at levels higher than 100 $\mu g\ g^{-1}$. Healthy-looking kernels of inoculated ears contained the much lower amounts of fumonisins compared to the FDK fraction, ranging from 0.06 to 1.5 $\mu g\ g^{-1}$ of FB_1 and from 0.02 to 0.4 $\mu g\ g^{-1}$ of FB_2 (data not shown). This clearly indicates that damaged kernels account for almost all the fumonisin contamination.

The totality of data, including all hybrids tested from one to four seasons, led to a good correlation between the *Fusarium* ear rot index (Fi) and fumonisin contamination, with a coefficient of correlation (r) of 0.933 (Pascale et al., 1997).

Inoculation with F. proliferatum. Fumonisin concentrations in maize kernels ranged from 18.0 to 231.9 $\mu g\ g^{-1}$ for FB_1 and from 0.4 to 26.5 $\mu g\ g^{-1}$ for FB_2, from 0.2 to 19.6 $\mu g\ g^{-1}$ for BEA and from 0.3 to 6.4 $\mu g\ g^{-1}$ for FP. Similarly to the disease score, a considerable influence of the year of inoculation on mycotoxin accumulation in maize hybrids was observed. In particular, significantly higher amounts of mycotoxins were detected in kernels of hybrids inoculated in the 1999 season, compared to kernels of the same hybrids inoculated in the 1996 and 1997 seasons. Between hybrids, comparison showed that Mona and RAH 895 contained low amounts of fumonisins, beauvericin and fusaproliferin in 1996, but were highly contaminated in 1997, whereas the hybrid KLG 2210 accumulated high concentrations of all four mycotoxins in 1996 and the lowest concentrations in 1997.

The amount of fumonisins in *Fusarium*-damaged kernels varied from 115.7 to 361.5 $\mu g\ g^{-1}$ and from 8.3 to 41.1 $\mu g\ g^{-1}$ for FB_1 and FB_2, respectively. The fraction of healthy-looking kernels of inoculated ears contained much lower levels of fumonisins compared to the *Fusarium*-damaged kernels fraction, ranging from 1.6 to 26.0 $\mu g\ g^{-1}$ of FB_1 and from 0.1 to 2.3 $\mu g\ g^{-1}$ of FB_2. The contamination levels of beauvericin and fusaproliferin in *Fusarium*-damaged kernels were significantly lower than fumonisins, ranging from 1.1 to 44.3 $\mu g\ g^{-1}$ for BEA and from 1.7 to 10 $\mu g\ g^{-1}$ for FP. Also, these toxins accumulated mainly in *Fusarium*-damaged kernels, whereas only trace amounts were found in healthy-looking kernels (up to 0.2 $\mu g\ g^{-1}$ for BEA and 0.3 $\mu g\ g^{-1}$ for FP). The analysis of control ears (not inoculated) showed that only four samples out of 30 contained fumonisin B_1 at levels not exceeding 1.2 $\mu g\ g^{-1}$, whereas fumonisin B_2, beauvericin and fusaproliferin were not detected (data not shown).

A good correlation between the *Fusarium* ear rot index (Fi) and mycotoxin contamination (FB_1, FB_2, BEA and FP) was observed (correlation coefficient (r) = 0.895). The hybrids Buran (in 1996) and Janina (in 1999) were not considered in the analysis. The high

Table 6. Meteorological data (mean monthly temperatures and monthly precipitations) for 1992–1997 and 1999 at Radzikow and long-term averages

Month	Mean temperatures (°C)								Precipitation sums (mm)							
	1992	1993	1994	1995	1996	1997	1999	Long-term 1966–1995	1992	1993	1994	1995	1996	1997	1999	Long-term 1966–1995
May	14.2	18.2	13.8	13.8	16.4	14.8	15.7	13.8	17.6	36.3	64.5	41.6	69.4	56.0	39.3	47.6
June	18.3	17.4	17.6	18.6	18.2	18.1	19.7	16.7	47.3	20.0	8.3	73.1	66.8	84.3	155.1	62.1
July	20.2	18.2	23.9	22.1	17.5	19.1	23.1	18.4	55.8	65.5	22.8	63.3	132.6	227.7	30.7	67.8
August	21.6	18.1	19.4	19.8	20.2	20.9	20.6	17.8	24.2	49.5	71.9	66.1	39.7	31.9	5.0	54.8
September	12.7	12.6	15.4	14.1	11.2	13.9	18.3	13.3	78.6	46.5	53.6	138.7	55.6	18.6	5.4	45.4
October	5.9	8.5	6.9	11.1	9.8	6.8	8.9	8.4	39.8	27.6	74.2	8.7	25.3	26.6	18.8	35.8

levels of mycotoxins found in kernels of these hybrids were not in agreement with the external symptoms of infection observed; this could be explained by a possible fungal infection acting mainly inside the kernels (Pascale et al., 2001).

Discussion

The results showed that *F. verticillioides* and *F. proliferatum* were highly pathogenic to maize ears under field conditions. These species may contribute to significant contamination of maize with at least four toxic metabolites (i.e. FB_1, FB_2, BEA and FP). Damaged kernels accounted for almost all the mycotoxin contamination. Consequently, it is possible to drastically reduce the exposure of animals and humans to mycotoxins by segregating mouldy and visibly damaged kernels from healthy kernels.

A good correlation between the *Fusarium* ear rot index (Fi) and mycotoxin contamination was observed, both in *F. verticillioides* and *F. proliferatum* inoculation experiments. The severity of ear infection has been also found to be a good indicator of ear contamination by mycotoxins (Schaafsma et al., 1993).

Poor reproducibility of the results was observed between the different years of investigation. Low disease scores and low content of mycotoxins, observed in one or two seasons, did not necessarily produce low susceptibility of a given hybrid in other seasons. The highest disease scores and accumulation of mycotoxins in the maize hybrids were observed in the years 1994 and 1995 for the *F. verticillioides* inoculation experiments, and in the year 1999 for the *F. proliferatum* inoculation experiments. These seasons were characterized by high temperatures at the pollination period (July), but low precipitation was observed only in 1994 and 1999, not in 1995 (see Table 6). High temperatures, low moisture and dry weather at or just prior to pollination have been reported to be important factors in fumonisin production (Shelby et al., 1994). In the USA, fumonisin contamination in warm, dry years was reported to be greater than in cooler years (Murphy et al., 1996).

Fumonisin levels in maize kernels after inoculation with *F. proliferatum* were much higher than those found in experiments carried out with *F. verticillioides*. Beauvericin levels in 1996 and 1997 were similar to those found in kernels of maize hybrids artificially inoculated with *F. subglutinans* (Krska et al., 1996), a known producer of beauvericin, whereas in 1999 they were much higher. No previous inoculation experiments under field conditions have been performed with respect to fusaproliferin accumulation, although high levels of the toxin have been found in pre-harvest maize ear rot (Ritieni et al., 1997b).

The ability of maize hybrids to accumulate only low amounts of mycotoxins is a highly desired trait for farmers. In our experiments, a generally high susceptibility to fungal infection and high accumulation of mycotoxins in kernels was observed. However, some promising traits were found in the hybrid Mona resulting in a low *F. verticillioides* ear rot index and low fumonisin contamination. These results suggest that the possibilty of finding selected hybrids with different susceptibility to *Fusarium* ear rot and different tendencies to accumulate mycotoxins in the field exists, but further long-term experiments are necessary before maize hybrids resistant to infection by *Fusarium* species and consequent mycotoxin accumulation are available commercially.

References

Bottalico A, Logrieco A and Visconti A (1989) *Fusarium* species and their mycotoxins in infected cereals in field and in stored grains. In: Chelkowski J (ed) *Fusarium*-Mycotoxins, Taxonomy and Pathogenicity. Topics in Secondary Metabolites (pp 85–119) Elsevier, Amsterdam, the Netherlands

Bullerman LB (1996) Occurrence of *Fusarium* and fumonisins in food grains and in foods. In: Jackson LS, De Vries JW and Bullermann LB (eds) Advances Experimental Medicine and Biology – Fumonisins in Food (pp 27–38) Plenum Press, New York, USA

Chelkowski J (1989) Mycotoxins associated with corn cob fusariosis. In: Chelkowski J (ed) *Fusarium*-Mycotoxins, Taxonomy, Pathogenicity. Topics in Secondary Metabolism (pp 53–62) Elsevier, Amsterdam, the Netherlands

Chulze SN, Ramirez ML, Pascale M and Visconti A (1998) Fumonisin production by, and mating population of, *Fusarium* section *Liseola* isolates from maize in Argentina. Mycological Research 102: 141–144

Grove JF and Pople M (1980) The insecticidal activity of beauvericin and the enniatin complex. Mycopathologia 70: 103–105

Hart LP, Braselton WE and Stebbins TC (1982) Production of zearalenone and deoxynivalenol in commercial sweet corn. Plant Disease 66: 1133–1135

IPCS (2000) Environmental Health Criteria 219 – Fumonisin B_1. International Programme on Chemical Safety: World Health Organization (WHO), Geneva

Krska R, Lemmens M, Schuchmacher R, Grasserbauer M, Pronczuk M, Wisniewska H and Chelkowski J (1996) Accumulation of the mycotoxin beauvericin in kernels of corn hybrids inoculated with *Fusarium subglutinans*. Journal of Agricultural and Food Chemistry 44: 3665–3667

Krska R, Schuchmacher R, Grasserbauer M, Lemmens M, Lemmens-Gruber R, Adler A and Lew H (1997) Effects of beauvericin to mammalian tissue and its production by Austrian isolates of *Fusarium proliferatum* and *Fusarium subglutinans*. Mycotoxin Research 13: 11–16

Logrieco A, Moretti A, Ritieni A, Chelkowski J, Altomare C, Bottalico A and Randazzo G (1993) Natural occurrence of beauvericin in pre-harvest *Fusarium subglutinans* infected corn ears in Poland. Journal of Agricultural and Food Chemistry 41: 2149–2152

Logrieco A, Moretti A, Ritieni A, Bottalico A and Corda P (1995) Occurrence and toxigenicity of *Fusarium proliferatum* from pre-harvest maize ear rot, and associated mycotoxins in Italy. Plant Disease 79: 727–731

Logrieco A, Moretti A, Fornelli F, Fogliano V, Ritieni A, Caiaffa M, Randazzo G, Bottalico A and Macchia L (1996) Fusaproliferin production by *Fusarium subglutinans* and its toxicity to *Artemia salina*, SF-9 insect cells and IARC/LCL 171 human B lymphocytes. Applied and Environmental Microbiology 62: 3378–3384

Logrieco A, Moretti A, Castella G, Kostecki M, Golinski P, Ritieni A and Chelkowski J (1998) Beauvericin production by *Fusarium* species. Applied and Environmental Microbiology 64: 3084–3088

Marasas WFO (1995) Fumonisins: their implications for human and animal health. Natural Toxins 3: 193–198

Moretti A, Logrieco A, Bottalico A, Ritieni A, Fogliano A and Randazzo G (1997) Diversity in beauvericin and fusaproliferin production by different populations of *Gibberella fujikuroi* (*Fusarium* section *Liseola*). Sydownia 48: 45–56

Munkvold G, Stahr HM, Logrieco A, Moretti A and Ritieni A (1998) Occurrence of fusaproliferin and beauvericin in *Fusarium*-contaminated livestock feed in Iowa. Applied and Environmental Microbiology 64: 3923–3926

Murphy PA, Hendrich S, Hopmans EC, Hauck CC, Lu Z, Buseman G and Munkvold G (1996) Effect of processing on fumonisins content of corn. In: Jackson LS, De Vries JW and Bullermann LB (eds) Advances Experimental Medicine and Biology – Fumonisins in Food (pp 232–334) New York, USA

Ojcius DM, Zychlinsky A, Zheng LM and Young JD (1991) Ionophore-induced apoptosis: Role of DNA fragmentation and calcium fluxes. Experimental Cell Research 197: 43–49

Pascale M, Visconti A, Pronczuk M, Wisniewska H and Chelkowski J (1997) Accumulation of fumonisins in maize hybrids inoculated under field conditions with *Fusarium moniliforme* Sheldon. Journal of the Science of Food and Agriculture 74: 1–6

Pascale M, Visconti A, Avantaggiato A, Pronczuk M and Chelkowski J (1999) Mycotoxin contamination of maize hybrids after infection with *Fusarium proliferatum*. Journal of the Science of Food and Agriculture 79: 2094–2098

Pascale M, Visconti A, Pronczuk M. Wisniewska H and Chelkowski J (2001) Accumulation of fumonisins, beauvericin and fusaproliferin in maize hybrids inoculated under field conditions with *Fusarium proliferatum*. Mycological Research (submitted)

Ritieni A, Monti SM, Randazzo G, Logrieco A, Moretti A, Peluso G and Fogliano V (1997a) Teratogenic effects of fusaproliferin on chicken embryos. Journal of Agricultural and Food Chemistry 45: 3039–3043

Ritieni A, Moretti A, Logrieco A, Bottalico A, Randazzo G, Monti SM, Ferracane R and Fogliano V (1997b) Occurrence of fusaproliferin, fumonisin B_1 and beauvericin in maize from Italy. Journal of Agricultural and Food Chemistry 45: 4011–4016

Schaafsma AW, Miller JD, Savard ME and Ewing RJ (1993) Ear rot development and mycotoxin production in corn in relation to inoculation method, corn hybrid and species of *Fusarium*. Canadian Journal of Plant Pathology 15: 185–192

Shelby RA, White DG and Basuke EM (1994) Differential fumonisin production in maize hybrids. Plant Disease 78: 582–584

Shephard GS, Sewram V, Nieuwoudt TW, Marasas WFO and Ritieni A (1999) Production of the mycotoxins fusaproliferin and beauvericin by South African isolates in the *Fusarium* section *Liseola*. Journal of Agricultural and Food Chemistry 47: 5111–5115

US NPT (1999) NPT Technical Report on the 'Toxicology and carcinogenesis studies of fumonisin B_1 (CAS No. 116355-83-0) in F344/N rats and B6C3F_1 mice (feed studies)'. NPT TR 496, NIH Publication No. 99-3955, US Department of Health and Human Services.

Visconti A (1996) Fumonisins in maize genotypes grown in various geographic areas. In: Jackson LS, De Vries JW and Bullermann LB (eds) Advances Experimental Medicine and Biology – Fumonisins in Food (pp 193–204) Plenum Press, New York, USA

European Journal of Plant Pathology **108:** 653–660, 2002.

Studies on the infection process of *Fusarium culmorum* in wheat spikes: Degradation of host cell wall components and localization of trichothecene toxins in infected tissue

Z. Kang and H. Buchenauer
Institute of Phytomedicine, University of Hohenheim, D-70593 Stuttgart, Germany
(Fax: +494592408; E-mail: hbuchen@uni-hohenheim.de)

Key words: Fusarium culmorum, Triticum aestivum, infection process, deoxynivalenol, cell wall degrading enzymes

Abstract

After single spikelet inoculation, the infection process of *Fusarium culmorum* and spread of fungal hyphae in the spike tissues were studied by scanning and transmission electron microscopy. While hyphal growth on outer surfaces of the spike was scanty and no successful penetration was observed, the fungus developed a dense mycelium on the inner surfaces and effectively invaded the lemma, glume, palea and ovary by penetration pegs. During the inter- and intracellular spreading of the fungus, marked alterations in the host tissues were observed, including degeneration of cytoplasm, cell organelles, and depositions of electron dense material between cell wall and plasmalemma. Ultrastructural studies revealed that host cell walls in proximity of the penetration peg and in contact with hyphae were less dense or transparent which suggested that cell wall degrading enzymes were involved in colonisation of host tissues by fungal hyphae. Enzyme- and immunogold-labelling investigations confirmed involvement of extracellular enzymes, that is cellulases, xylanases and pectinases, in degradation of cell wall components. Localization studies of trichothecenes indicated that toxins could be detected in host tissues at an early stage of infection.

Introduction

Fusarium head blight or scab of wheat (*Triticum aestivum*) is one of the most destructive diseases in humid and semi-humid areas. The disease is caused by a number of *Fusarium* species, mainly *F. graminearum* (teleomorph *Gibberella zeae*), *F. culmorum* and *F. avenaceum* (teleomorph *G. avenacea*) (Bai and Shaner, 1994; Parry et al., 1995). The disease causes substantial yield losses as a result of a diminished number of grains per spike and reduced grain weight. The infected grains are of low quality with damaged starch granules and storage proteins (Bechtel et al., 1985).

Moreover, *Fusarium* species produce highly toxic metabolites, for example trichothecene mycotoxins, such as deoxynivalenol (DON), 3-acetyldeoxynivalenol (3-ADON) and 15-acetyldeoxynivalenol (15-ADON) (Wang and Miller, 1988; Snijders, 1990). These toxins are detrimental to livestock and human health (Trenholm et al., 1983; Marasas et al., 1984). In addition, some mycotoxins, for example DON, exert phytotoxic effects and are involved, as virulence factors, in the pathogenicity of *F. culmorum* and *F. graminearum* (Miller, 1989; Snijders and Krechting, 1992).

Wheat heads are most susceptible to the infection of the pathogen at the flowering stage (Pugh et al., 1933; Andersen, 1948; Strange and Smith, 1971). Detailed studies on the infection process of the pathogen on wheat heads, especially during the early stages, are limited and contradictory. In this study, the infection course of *F. culmorum* in wheat spikes was investigated by scanning and transmission electron microscopy. Also, the degradation of host cell wall components in *F. culmorum*-infected wheat spikes was demonstrated by means of enzyme-gold and immunogold-labelling.

Finally, the distribution of *Fusarium* toxins in wheat spikes at early infection stages was revealed by immunogold-labelling.

Materials and methods

Wheat plants of the cultivar Agent, which is susceptible to *Fusarium* head blight, were grown in 12 litre pots under outdoor conditions and inoculated at mid-flowering (GS 65) with a spore suspension of *F. culmorum* (isolate no. 46; Institute of Phytomedicine) by spraying the whole wheat spikes with a macroconidial suspension (5×10^5 spores ml^{-1}) or by injecting a conidial suspension (10^5 spores ml^{-1}) into the floral cavity between the lemma and palea of the first floret on spikelets using a pipette. Control plants were treated with distilled water. Immediately after inoculation, spikes were covered with plastic bags and plants were transferred into a growth chamber (16 h fluorescent light at 23 °C and 8 h darkness at 18 °C). After two days, the plastic bags were removed and plants were returned to outdoor conditions.

Whole single spikelets, including glumes, lemmas, paleas, ovaries and rachis from inoculated wheat spikes were sampled 6 and 12 h and 1, 2, 3, 4, 6 and 8 days after inoculation. The samples were fixed in glutaraldehyde (3–4%, v/v) in 50 mM phosphate buffer for 3–8 h at 4 °C. The samples were processed for scanning electron microscopy and light and transmission electron microscopy (Kang and Buchenauer, 2000a). The mycotoxins deoxynivalenol (DON) and 3-acetyldeoxynivalenol (3-ADON) produced by *F. culmorum* were localized in host and fungal cells by immunogold-labelling (Kang and Buchenauer, 1999). The cell wall components cellulose, xylan and pectin were detected in healthy and *F. culmorum*-infected wheat spikes using enzyme-gold and immunogold-labelling methods (Kang and Buchenauer, 2000b).

Results

F. culmorum macroconidia germinated with several germ tubes within 6–12 h after inoculation. Further development of the germ tubes differed markedly on the outer and inner surfaces of the floret tissues.

Growth of the hyphae on the outer surfaces was severely retarded. Two days after inoculation, the germ tubes on the convex middle part of the glume were short and possessed morphological alterations. Four days after inoculation, hyphal growth was very limited and no hyphal network developed. Hyphae near the edges of the lemma and glume exhibited more intensive growth, tending to extend to the edges in order to reach the inner surfaces of the glume, lemma or palea (Figure 2). Two days after inoculation, the fungus had developed a dense hyphal network on the inner surfaces of the lemma (Figure 1), glume, palea and top of the ovary. A dense mycelium developed on the anthers and pollen grains which were often present between the brush hair on the top of the ovary. With extended incubation time, hyphal growth on the inner surfaces of the spikelet became more dense and 3–4 days after inoculation the hyphae had reached the bases of the lemma, glume and palea. *F. culmorum* did not penetrate the inner surfaces of the host tissue immediately after germination of the macroconidia, but infections were observed after development of the dense hyphal network.

By 36 h after inoculation, hyphae in contact with the inner surfaces of the spikelet tissues formed short branches, the infection hyphae. The hyphae and the infection hyphae in touch with the host surfaces often produced cutinases and the cuticular layer beneath the hyphae was digested. The infection hyphae did not form appressoria but penetrated the epidermal cell wall directly by developing a constricted penetration peg. A septum was usually formed between the hyphal cell and the penetration peg. The fungus did not always penetrate the epidermal cell wall completely, but sometimes hyphae developed between the cuticule and the cell wall or within the cell wall layers. After penetration, during the first phase of tissue colonization, hyphae predominantly extended intercellularly between parenchyma cells in the lemma (Figure 3) and palea. During growth in the intercellular spaces, the fungus induced alterations in the host cells which were in contact or in advance of the hyphae. Small appositions were detected between the secondary cell wall and the plasmalemma. In the vacuoles, electron dense material was found and chloroplasts became swollen. Subsequently, the pathogen penetrated the affected host cells and colonized the tissues of the lemma, palea and ovary by both inter- and intracellular hyphal growth.

Three to four days after inoculation, the hyphae of *F. culmorum* had reached the bases of the lemma and palea by growing on the inner surfaces of the lemma and palea. At that time, the fungus had spread inter- and intracellularly to the base of the lemma and palea as well as horizontally from the inner surfaces of the lemma and palea to their outer surfaces. Typical

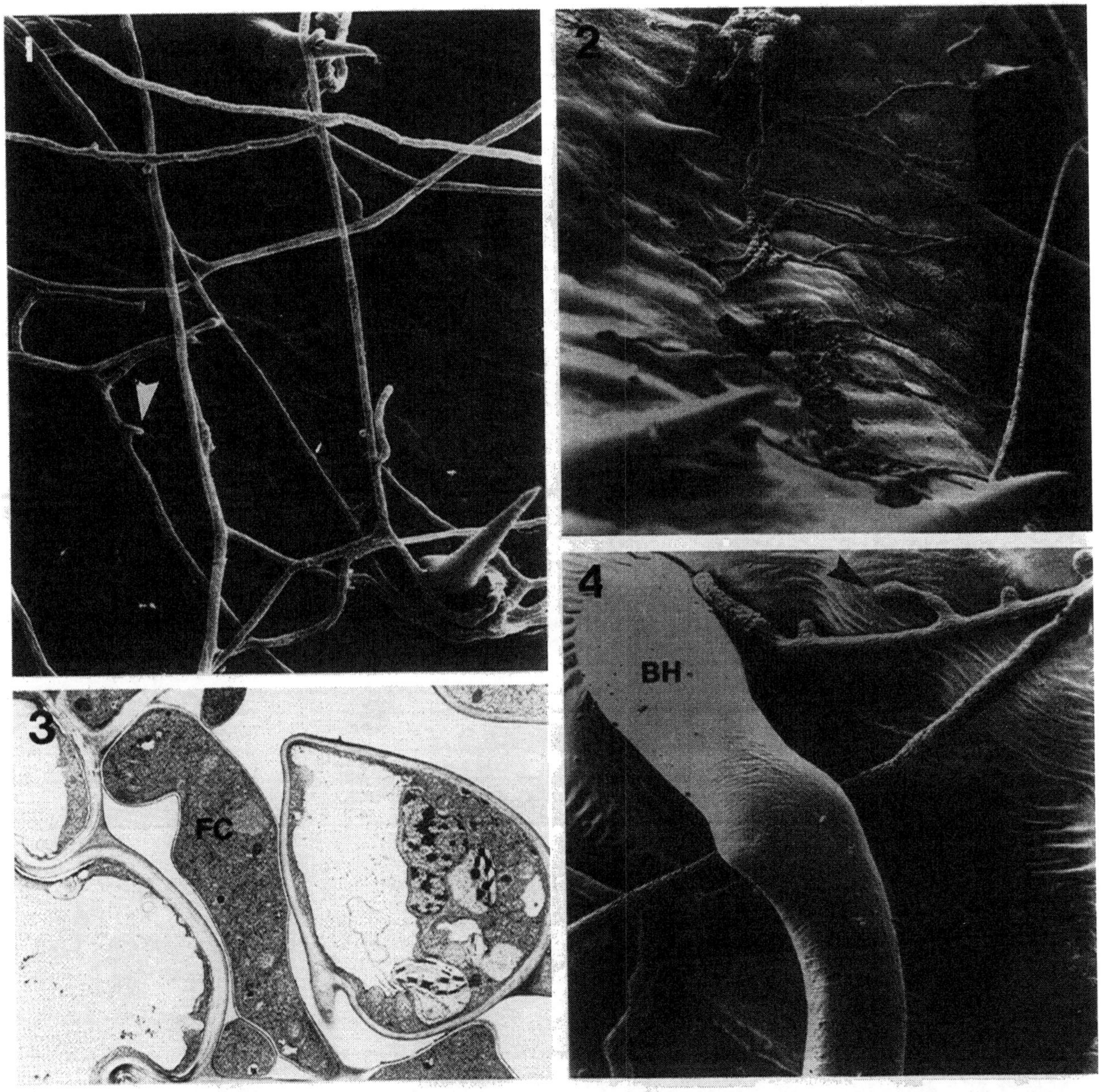

Figures 1–4. Scanning and transmission electron micrographs of colonization of wheat spike by *Fusarium culmorum*. **1**: Hyphal network on the inner surface of the lemma, 2 days after inoculation (dai). Infection hyphae (arrow) were already formed from hyphae adhering to the surface of the lemma. **2**: Hyphae on the outer surface near the top edge of the glume, 3 dai. The hyphae grew upwards and over the edge of the glume. **3**: Hyphae in the intercellular spaces between parenchyma cells in the lemma, 2 dai. Appositions appeared between host cell wall and plasmalemma. **4**: Hyphae with infection hyphae (arrow) on the top of the ovary, 2 dai. FC: fungal cell; BH: brush hair.

symptoms of brown water-soaked spots appeared on the infected tissues.

The cells of the colonized host tissue exhibited severe damage, such as disintegration of the cytoplasm and cell organelles, as well as collapse of some parenchyma cells. Host cell walls in close contact with hyphae or around the penetration pegs were, in general, less dense or transparent, suggesting that the pathogen produced cell-wall degrading enzymes during colonization of the host tissue.

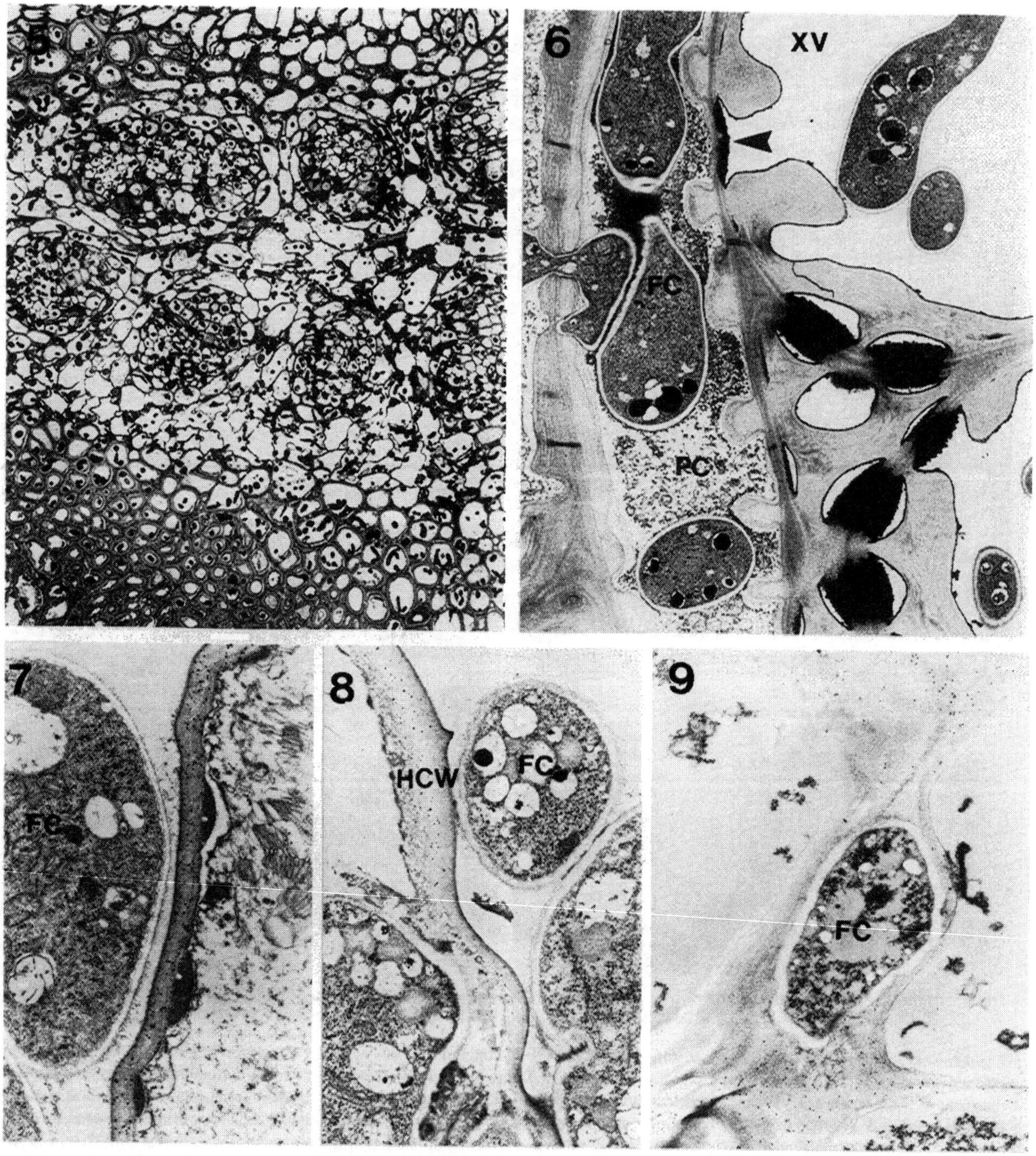

Figures 5–9. Light and transmission electron micrographs of colonization of wheat spike by *Fusarium culmorum*. **5**: A cross-section of the rachis node, 4 dai. Hyphae are distributed in the vascular bundles, the parenchyma tissue outside the vascular bundle, and also in the thick-walled tissue around the vascular tissue. **6**: A cross-section of the rachis, 6 dai. Hyphae spreading in the xylem vessels and the paratracheal parenchyma cells. The primary wall of vessels became much more dense and there was a dense coating on the inner surfaces of the vessel walls (arrow). **7**: Cytochemical labelling for cellulose by a cellulase-gold probe over the cell walls in the *Fusarium culmorum*-infected lemma, 3 dai. The density of gold particles over the host cell wall decreased, particularly adjacent to the hypha. No labelling was seen over the appositions between the host cell wall and plasmalemma. The fungal cell wall was free of gold particles.

Fungal hyphae growing on the upper surface of the ovary also developed infection hyphae about 36 h after inoculation (Figure 4) and usually penetrated the tissue through the junctions between the epidermis or between the epidermis and the brush hair. The hyphae either extended between the epidermal wall layers or penetrated the cell wall. The fungus also infected the stigma and grew downwards to the upper pericarp cells. Hyphae extended very rapidly by inter- and intracellular growth from the upper part of the pericarp to the base of the ovary. The pericarp cells invaded by the fungus or in proximity to the intercellular hyphae were degenerated and host cell walls had disintegrated.

The hyphae extending either on the inner surface of the lemma, palea and the outer surface of the ovary or spreading within the host tissues reached the base of the lemma, palea or ovary about 3–4 days after inoculation and entered the rachilla, which is characterized by a central core of very thick-walled cells. After colonization of the rachilla, hyphae extended into the rachis node (Figure 5) and then grew into the rachis, which was reached 4–5 days after inoculation. In the rachis, hyphae of *F. culmorum* spread in the xylem vessels, phloem sieves and paratracheal parenchyma cells and colonized the rachis by inter- and intracellular hyphal extension (Figure 6). *F. culmorum* started to produce conidiophores and macroconidia when the outer surface of the lemma and palea had been reached, 5–6 days after inoculation.

The cell wall components cellulose, xylan and pectin were localized in different tissues of the non-inoculated, healthy and *F. culmorum*-infected wheat spikes by means of enzyme-gold and immuno-gold labelling (Kang and Buchenauer, 2000b). The intra- and intercellular colonization of the lemma, palea, ovary and rachis of the wheat spike by hyphae of *F. culmorum* caused pronounced alterations of the matrices of the host cell wall and middle lamella, and also resulted in marked modifications of the host cell wall components such as cellulose, xylan and pectin (Figures 7–9). The significant reduced binding sites of the cell wall macromolecules revealed by enzyme gold and immunogold-labelling of the infected tissue, as compared to the healthy non-infected cell walls, indicated excretion of cellulases, xylanases and pectinases by the hyphae of the pathogen during penetration and spreading. From the results obtained, it may be deduced that during the initial infection and spreading process of the pathogen, pectinases activity is higher than the activities of the cellulases and xylanases.

An immunogold-labelling method was also developed to study the subcellular localization of the trichothecene mycotoxins deoxynivalenol (DON) and 3-acetyldeoxynivalenol (3-ADON) (Kang and Buchenauer, 1999). The toxins were produced by the hyphae before penetration, when they were growing in close contact on the inner surface of the lemma and palea or on the surface of the ovary. Thirty-six hours after inoculation, toxins were detected in the walls and cells of the epidermis. During penetration, toxins were found in the host cell wall around the infection peg and in the cytoplasm of the host cells (Figure 10). During spreading of the pathogen, higher intensities of toxin labelling were present in host cells in proximity to the hyphae.

In plant cells, toxins were localized in the cytoplasm, chloroplasts, plasmalemma and sometimes associated with endoplasmic reticulum and ribosomes (Figure 11). Figure 12 shows a parenchyma cell of the lemma with disorganized cytoplasm, 3 dai. Toxins were present over the host cell wall, cytoplasm and vacuole.

In infected wheat kernels, toxins were detected at different concentrations in the pericarp tissues, pigment strand, aleurone cells and in the starch endosperm. In the rachis colonized by the fungus, toxins were found in xylem vessels (Figure 13) and phloem sieve tubes and parenchyma cells outside the vascular bundles.

The studies also revealed that toxins were translocated upwards in the xylem and phloem sieve tubes and downwards in the phloem sieve tubes to areas of the spike which were not colonized by the fungus.

Discussion

The study revealed rapid and effective colonization of wheat spikes by *F. culmorum*. The hyphae of the pathogen extended inter- and intracellularly in the parenchyma and vascular tissues of the lemma, glume,

8: Cytochemical localization of xylan with a xylanase-gold probe in cell walls of the *Fusarium culmorum*-infected ovary, 3 dai. The cytoplasm of the cells was destroyed and labelling decreased over the host cell wall in contact with the hyphae. **9**: Immunogold labelling for pectin by monoclonal antibody, JIM7, in the cell walls of the *Fusarium culmorum*-infected sieve tubes in the rachis, 5 dai. The growth of the hypha in the middle lamella of the sieve tube resulted in cell wall displacement and reduction of gold particles in the wall. FC: fungal cell; HCW: host cell wall; PC: paratracheal parenchyma cell; XV: xylem vessel.

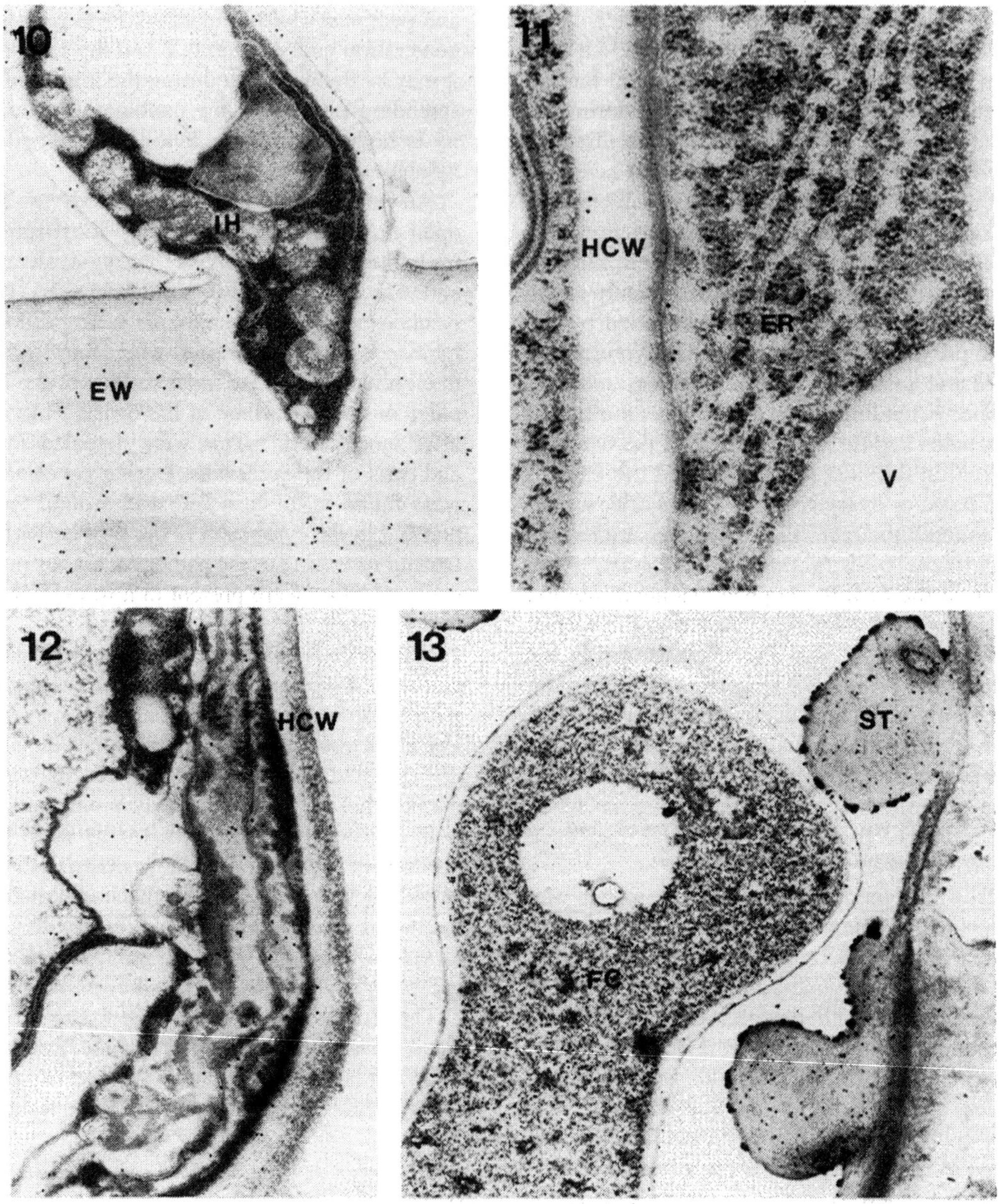

Figures 10–13. Immunogold localization of Fusarium toxins in infected host tissue by *Fusarium culmorum*. **10**: The epidermal cell wall of the lemma, 36 h after inoculation, was incubated with DON-antiserum and goat anti-rabbit immunoglobulin-linked colloidal gold particles (GAR-gold antibody). Gold particles were localized over the host cell wall and the hyphal cell wall. Note that dense accumulation of gold particles was found over the host cell wall in advance of the penetration peg. **11**: The epidermal cell in the ovary, 2 dai, incubated with DON-antiserum and GAR-gold antibody. The gold particles were found over the cell wall, cytoplasm, endoplasmic reticulum and ribosomes. **12**: The parenchyma cell with disorganized cytoplasm in the lemma, 3 dai, incubated with DON-antiserum and GAR-gold antibody. Gold labelling was found over the host cell wall, cytoplasm and the vacuole. **13**: The infected xylem vessel in the rachis, 6 dai, incubated with DON-antiserum and GAR-gold antibody. The vessel was labelled with numerous gold particles on the secondary thickenings and with fewer gold particles on the primary wall. The hyphal cell in the vessel was also labelled. ER: endoplasmic reticulum; EW: epidermal cell wall; FC: fungal cell; HCW: host cell wall; IH: infection hypha; ST: secondary thickening; V: vacuole.

ovary and rachis, and caused severe damage to the host tissues.

Scanning microscopical studies revealed striking differences regarding hyphal growth of *F. culmorum* on the outer surfaces of the glume, lemma and rachis and the inner surfaces of the lemma, glume and palea. While the fungus developed a dense mycelium on the inner surfaces and effectively invaded these tissues, hyphal growth on outer surfaces of the spike was scanty and no successful penetration was observed.

The unfavourable conditions for hyphal growth on the outer surfaces of the wheat spikes may be associated with physical and chemical as well as environmental properties. The pronounced wax layer covered with wax crystals may present a water-repellent surface and also might contain antifungal components. As a result of these inhospitable conditions, almost no infections by *F. culmorum* occurred on the outer surfaces. These observations are in agreement with studies of Strange and Smith (1971) which showed that no infections occurred after placing droplets of spore suspensions on the outer surface of the glume.

Ultrastructural analysis showed that the host cell walls in contact with the hyphae or in proximity to the penetration pegs were less dense or transparent suggesting that cell-wall degrading enzymes are involved in the spread of the fungus in the host tissues. Cytochemical labelling studies confirmed degradation of the cell wall components cellulose, xylan and pectin in *F. culmorum*-infected wheat spikes. The study also showed that during the early stage of infection (e.g. 3 dpi) pectin degradation in the epidermal cell walls of the wheat ovary was more pronounced than that of cellulose and xylan. This finding suggests that *F. culmorum* may secrete pectinases earlier or in higher activities than cellulases or xylanases. Following infection, the pathogen invaded host cells only when they exhibited signs of disintegration or necrosis.

Numerous studies have shown that *F. culmorum* produces the trichothecene mycotoxins DON, 3-ADON, 15-ADON and nivalenol (Wang and Miller, 1988; Wong et al., 1995). It has been demonstrated that these toxins also display detrimental effects to plant cells such as inhibition of protein synthesis, electrolyte losses, cytoplasm convolution and disintegration of organelles (Miller and Ewen, 1997; Pavlovkin et al., 1986).

The studies indicated, that as early as 36 h after inoculation, hyphae of *F. culmorum* growing on the surfaces of the lemma and ovary had already secreted trichothecene toxins which were present in the host tissues before the pathogen had invaded the host cells. During penetration and growth the hyphae continued to produce toxins which were detected in host cells adjacent to and in advance of the hyphae. In the host cells, *Fusarium* toxins were associated with the plasmalemma and in the cytoplasm sometimes with ribosomes and endoplasmic reticulum. These localization studies suggest that the trichothecene toxins interfere with the wheat spike/*F. culmorum*-interaction by impairing host defence reactions. Their association with the plasmalemma might result in alterations of membrane permeability, and because of the affinity to the ribosomes, it is assumed that the toxins interfere in protein synthesis which might result in severe impairment of post-infection defence mechanisms. The delayed or inhibited post-infection responses also became evident from the ultrastructural studies. It was difficult to detect defence structures such as callose or papilla formations in the infected tissues.

Wheat kernels colonized by *F. culmorum* were also contaminated with the toxins (Snijders and Perkowski, 1990; Sinha and Savard, 1997). The labelling density in the different tissues not only correlated with the hyphal density; toxins were also detected in advance of the hyphae in the starchy endosperm. It may be concluded that, in addition to the production of cell-wall degrading enzymes, the secretion of trichothecene mycotoxins may also constitute an important factor in head blight development of wheat by *F. culmorum*.

Acknowledgements

The authors thank Mrs H. Brandl and Mrs M. Harfold for excellent technical assistance, Mrs G. Moll for preparing the photographs and Mrs M. Sarkar for typing the manuscript. This study has been supported by the Deutsche Forschungsgemeinschaft (DFG; Forschergruppe 'Fusarium-Toxine').

References

Anderson AL (1948) The development of *Gibberella zeae* head blight of wheat. Phytopathology 38: 595–611

Bai G and Shaner G (1994) Scab of wheat: Prospects for control. Plant Disease 78: 760–766

Bechtel DB, Kaleikan LA, Gaines RL and Seitz LM (1985) The effects of *Fusarium graminearum* infection on wheat kernels. Cereal Chemistry 62: 191–197

Kang Z and Buchenauer H (1999) Immunocytochemical localization of Fusarium toxins in infected wheat spikes by *Fusarium culmorum*. Physiological and Molecular Plant Pathology 55: 275–288

Kang Z and Buchenauer H (2000a) Cytology and ultrastructure of the infection of wheat spikes by *Fusarium culmorum*. Mycological Research 104: 1083–1093

Kang Z and Buchenauer H (2000b) Ultrastructural and cytochemical studies on cellulose, xylan and pectin degradation in wheat spikes infected by *Fusarium culmorum*. Journal of Phytopathology 148: 263–275

Marasas WFO, Nelson PE and Toussoun TA (1984) Toxigenic *Fusarium* species: Identity and mycotoxicology. The Pennsylvania State University Press, University Park, PA

Miller JD (1989) Effects of *Fusarium graminearum* metabolites on wheat cells. In: Graniti A, Durbin RD and Ballio A (eds) Phytotoxins and Plant Pathogenesis. NATO ASI Ser. Ser. H, Cell Biology 27: 449–452

Miller JD and Ewen MA (1997) Toxic effects of deoxynivalenol on ribosomes and tissues of the spring wheat cultivars Frontana and Casavant. Natural Toxins 5: 234–237

Parry DW, Jenkinson P and Mc Leod L (1995) *Fusarium* ear blight (scab) in small grain cereals – a review. Plant Pathology 44: 207–238

Pavlovkin J, Mistrik I, Zajcenko AM and Mraz J (1986) Effects of mycotoxins on cell membrane of higher plant roots. Biologia (Bratislava) 41: 681–686

Pugh GW, Johann H and Dickson JG (1933) Factors affecting infection of wheat heads by *Gibberella sambinetii*. Journal of Agricultural Research 46: 771–791

Sinha RC and Savard ME (1997) Concentration of deoxynivalenol in single kernels and various tissue of wheat heads. Canadian Journal of Plant Pathology 19: 8–12

Snijders CHA and Krechting CF (1992) Inhibition of deoxynivalenol translocation in *Fusarium* head blight resistant wheat. Canadian Journal of Botany 70: 1570–1576

Snijders CHA and Perkowski J (1990) Effects of head blight caused by *Fusarium culmorum* on toxin content and weight of wheat kernels. Phytopathology 80: 566–570

Snijders CHA (1990) *Fusarium* head blight and mycotoxin of wheat, a review. Netherlands Journal of Plant Pathology 96: 187–198

Strange RN and Smith H (1971) A fungal growth stimulant in anthers which predisposes wheat to attack by *Fusarium graminearum*. Physiological Plant Pathology 1: 141–150

Trenholm HL, CochraneWP, Cohen H, Elliot JI, Farnworth ER, Friend DW, Hamilton RMG, Standish JF and Thompson BK (1983) Survey of vomitoxin contamination of the Ontario 1980 white winter wheat crop: Results of survey and feeding trials. Journal of the Association of Official Analytical Chemistry 66: 92–97

Wang YZ and Miller JD (1988) Effects of *Fusarium graminearum* metabolites on wheat tissue in relation to Fusarium head blight resistance. Journal of Phytopathology 122: 118–125

Wong LSL, Abramson D, Tekauz A, Leisle D and McKenzie RIH (1995) Pathogenicity and mycotoxin production of *Fusarium* species causing head blight in wheat cultivars varying in resistance. Canadian Journal of Plant Sciences 75: 261–267

European Journal of Plant Pathology **108:** 661–666, 2002.

Production of beauvericin by different races of *Fusarium oxysporum f. sp. melonis*, the Fusarium wilt agent of muskmelon

A. Moretti[1], A. Belisario[2], A. Tafuri[3], A. Ritieni[3], L. Corazza[2] and A. Logrieco[1]
[1]*Institute of Sciences of Food Production, C.N.R., Viale Einaudi 51, 70125, Bari, Italy (Fax: +390805486063; E-mail: moretti@area.ba.cnr.it);* [2]*Istituto Sperimentale per la Patologia Vegetale, Via C.G. Bertero 22, 00156 Roma, Italy;* [3]*Dipartimento di Scienza degli Alimenti, Università degli Studi di Napoli "Federico II", Portici 80055, Italy*

Key words: pathogenicity, races, melon, enniatins

Abstract

Fourty-four strains of *Fusarium oxysporum* were isolated from plants of melon with Fusarium wilt symptoms. Among these strains, thirty-nine were characterized for their pathogenicity on melon. Thirty-seven strains belonged to known races of *F. oxysporum f. sp. melonis*, while two strains were non-pathogenic. Four strains belonged to race 0, seven to race 1, four to race 2, and twenty-two to race 1,2. Beauvericin was produced by thirty-six strains in a range from 1 to 310 $\mu g\, g^{-1}$. Eight isolates of race 1,2 did not produce the toxin. In addition, of the two non-pathogenic strains, only one strain produced the toxin (290 $\mu g\, g^{-1}$). The production of enniatin A_1, enniatin B_1, and enniatin B was also investigated. Enniatin B was the only enniatin detected, being produced by eleven strains belonging to all the races, with a range of production from traces to 60 $\mu g\, g^{-1}$. Finally, melon fruits belonging to two different cultivars (*Cantalupo* and *Amarillo*) were artificially inoculated with one strain of *F. oxysporum f. sp. melonis* (ITEM 3464). Beauvericin was detected in the fruit tissues of both cultivars at a level of 11.2 and 73.8 $\mu g\, g^{-1}$, respectively. These data suggest that the production of both the toxins is not related to the pathogenicity of *F. oxysporum f. sp. melonis*, nor to the differential specificity of the races. The results confirm that beauvericin is a common metabolite of phytopathogenic *Fusarium* species.

Introduction

Beauvericin is a cyclodepsipeptide metabolite, closely related to the enniatins (Gaumann et al., 1960), a group of compounds with well known phytotoxic activity (Hermann et al., 1996; Burmeister and Plattner, 1987). Beauvericin was first reported as a compound produced by entomopathogenic fungi (Hamill et al., 1969; Bernardini et al., 1975; Gupta et al., 1991), and subsequently was shown to be produced by phytopathogenic fungi (Logrieco et al., 1993). In particular, beauvericin was found as a compound produced by strains of *Fusarium subglutinans* isolated from maize ears worldwide (Moretti et al., 1995). The role of beauvericin as an important secondary metabolite of phytopathogenic species of *Fusarium* was confirmed when production was detected in several strains of *F. proliferatum* isolated from different diseased host plants, including maize, asparagus, and date palm (Plattner and Nelson, 1994; Moretti et al., 1996; Abdalla et al., 2000). An investigation of the *Fusarium* genus, showed that various species produced beauvericin, including some strains of *F. oxysporum* from maize (Logrieco et al., 1998). Sagakuchi et al. (2000) provided the only available data on the phytotoxic effects of beauvericin by comparing it with other toxins (e.g. fusaric acid and fumonisin B_1). Although beauvericin did not cause any symptoms on roots of the plants tested (melon, tomato, wheat, and barley), it showed the highest toxicity towards all the protoplasts of the plants tested, and in particular towards melon protoplasts at the lowest concentration tested (10^{-2} μM; Sagakuchi et al., 2000). These findings opened new fields of investigation on the role that this metabolite could play in the muskmelon wilt disease.

Fusarium oxysporum f. sp. melonis is the causal agent of Fusarium wilt of muskmelon (*Cucumis melo*). Based on the pathogenicity reaction of a set of differential cultivars within *C. melo*, this *forma specialis* has been divided in to four races: 0, 1, 2, and 1,2, that correspond to the resistance genes that are overcome (Risser et al., 1976). Race 1,2 has been further subdivided into two different pathotypes 1,2y (yellows) and 1,2w (wilt), based on the type of symptoms induced (Bouhot, 1981). Although host specialisation has been historically identified as the most important trait in *F. oxysporum* and has been used as the basis of classification, the occurrence of multiple races in *F. oxysporum f. sp. melonis* showing diversity in virulence led scientists to question whether pathogenicity is the most meaningful basis on which to classify sub-*formae speciales* groupings (Jacobson and Gordon, 1991). The identification of eight vegetative compatibility groups (VCGs) within *F. oxysporum f. sp. melonis* and the fact that all four races can be in more than one VCG, as well as all the races being present in a single VCG (Jacobson and Gordon, 1990), showed that no race constitutes a genetically homogeneous group (Jacobson and Gordon, 1991). Moreover, races were not distinguished by a significant intergenic spacer (IGS) of the rDNA (Appel and Gordon, 1996). The physiological, genetic, or molecular determinants that could be the basis of pathogenic race differentiation are still unclear. Therefore, on the basis of this knowledge, the main object of this work was to investigate the production of beauvericin and related compounds as possible sources of biochemical determinants of pathogenicity to provide a better understanding of the relationships within this *formae speciales* of *F. oxysporum* and to assess the possible role that these metabolites could play in Fusarium wilt of muskmelon. Moreover, infection by *F. oxysporum f. sp. melonis* can also develop on fruits both in the field and during storage (Brigati and Gori, 1986). Therefore, some melon fruits were inoculated with a strain of *F. oxysporum f. sp. melonis* and the beauvericin occurrence in the fruits studied, in order to obtain information on the ability of this fungus to produce the toxin *in vivo*.

Materials and methods

Fungal strains

All fourty-four strains were isolated from muskmelon plants with typical Fusarium wilt symptoms. Thirty-two strains of *F. oxysporum f. sp. melonis* were isolated in the main areas of muskmelon cultivation in Italy by A. Belisario and L. Corazza (Istituto Sperimentale per la Patologia Vegetale (ISPaVe), Rome) and twelve strains were provided by T.R. Gordon (University of Davis, California) (Table 1). Tissues of muskmelon with symptoms of disease, were surface-disinfected with 1% NaOCl for 10 min, rinsed in sterile distilled water (SDW) and directly placed on Petri dishes containing a modified pentachloronitrobenzene medium selective for *Fusarium* (Nelson et al., 1983). Single spores of putative *Fusarium* colonies were transferred to carnation leaf agar (Nelson et al., 1983) for morphological identification.

Pathogenicity test

Fourteen-day-old cultures of *F. oxysporum* isolated from melon grown at room temperature (20–25 °C), were flooded with SDW and scraped with a sterile glass rod to obtain a purple–pink spore suspension. This slurry was filtered through two layers of sterile cheesecloth and the filtrate diluted with SDW to obtain spores (a mixture of macroconidia and microconidia) at a concentration of 10^6 ml^{-1}. Seeds of melon were surface-disinfected with 1% NaOCl for 20 min, rinsed in SDW, and sown in cell-type plastic growing trays (Bamaplast, Pistoia, Italy), one seed per cell (10 cm diam), filled with an autoclaved potting mix of peat and sand (1 : 1, v/v).

The differential varieties of muskmelon used were: Charentais T which lacks any genes for resistance, Charentais *Fom-1* resistant to race 0 and race 2, Charentais *Fom-2* resistant to race 0 and race 1, and Margot which carries both the resistance genes *Fom-1* and *Fom-2*. Seedlings were inoculated at the cotyledon to first-true-leaf stage when plants were about 10 days old, as this stage of maturity was shown to be most susceptible to Fusarium wilt. Plants were removed from the seedling trays and the roots washed in tap water, pruned to a length of about 2.5 cm, and dipped at the same time for 1 min into the spore suspension. The inoculated seedlings were transplanted into

Table 1. Origin, race, and toxin production of strains of *F. oxysporum f. sp. melonis*

Isolate ITEM	Isolate ISPaVE	Origin	Race	BEA ($\mu g\, g^{-1}$)*	ENN-B ($\mu g\, g^{-1}$)
3451	1279	Southern Italy	0	115	n.d
3463	1413	California	0	2	10
3465	1409	California	0	20	10
3471	1405	California	0	105	60
3281	1069	Sicily	1	100	n.d
3285	1074	Sicily	1	10	n.d
3288	1077	Sicily	1	75	n.d
3452	1281	Southern Italy	1	30	n.d
3464	1404	California	1	80	n.d.
3467	1407	California	1	35	15
3472	1418	California	1	1	n.d
3455	1286	Sicily	2	135	40
3461	1411	California	2	40	n.d.
3462	1412	California	2	15	10
3468	1406	California	2	10	3
3470	1417	California	2	2	Traces
3278	1018	Central Italy	1,2	90	10
3279	1027	Sicily	1,2	5	5
3280	1068	Sicily	1,2	65	40
3282	1070	Sicily	1,2	15	n.d
3283	1071	Sicily	1,2	n.d	n.d
3284	1073	Sicily	1,2	40	n.d
3286	1075	Sicily	1,2	n.d	n.d
3287	1076	Sicily	1,2	n.d	n.d
3289	1078	Southern Italy	1,2	265	n.d
3290	1079	Sicily	1,2	60	n.d
3291	1080	Sicily	1,2	310	n.d
3292	1081	Sicily	1,2	60	n.d
3293	1083	Southern Italy	1,2	n.d	n.d
3294	1084	Southern Italy	1,2	20	n.d
3295	1086	Southern Italy	1,2	25	n.d
3296	1299	Central Italy	1,2	30	n.d
3297	1222	Central Italy	1,2	35	n.d
3453	1291	Sicily	1,2	n.d	n.d
3456	1287	Southern Italy	1,2	5	n.d
3466	1415	California	1,2	n.d	n.d
3469	1416	California	1,2	95	n.d
3298	1296	Sicily	Nonpathogenic	n.d	n.d
3450	1304	Sicily	Nonpathogenic	290	n.d
3299	1219	Central Italy	—	140	n.d
3454	1299	Southern Italy	—	150	n.d
3300	1289	Southern Italy	—	105	20
3459	1303	Southern Italy	—	50	n.d
3460	1283	Sicily	—	55	n.d

ITEM: Istituto Tossine e Micotossine da Parassiti Vegetali, Bari, Italy; ISPaVE: Istituto Sperimentale per la Patologia Vegetale, Roma, Italy; BEA: beauvericin; ENN-B: enniatin B; n.d.: not detected; —: race not determined. *$\mu g\, g^{-1}$ of dried weight of fungal cultures.

plastic growing trays (one plant per cell) filled with a steamed potting mix (1 part each of soil : sand : peat) and placed in a greenhouse at 20/28 °C, night/day temperatures. The roots of control plants were pruned to a length of about 2.5 cm and dipped in tap water prior to transplanting. The number of dead and healthy plants was recorded 21 days after inoculation. Plants were rated as susceptible to Fusarium wilt if they died, or resistant to Fusarium wilt if they remained free of wilt symptoms.

In vitro *toxin production*

Single-conidium strains were grown on 10 g of autoclaved rice kernels adjusted to about 45% moisture in 50-ml Falcon tubes and inoculated with 2 ml of an aqueous suspension containing approximately 10^7 conidia ml^{-1}. Cultures were incubated at 25 °C for 4 weeks. The culture material was dried in a forced draft oven at 60 °C for 48 h, finely ground and stored at 4 °C. Controls were subjected to the same treatment but inoculated with SDW. Results were expressed at $\mu g\ g^{-1}$ dry weight.

In vivo *toxin production*

Melon fruits of two different cultivars (three for each cultivar), Cantalupo and Amarillo, were inoculated in the peduncle area with 2 ml of an aqueous suspension containing approximately 10^7 conidia ml^{-1}. The melon fruits were maintained at room temperature for one week. The fruit tissues were dried in a forced draft oven at 60 °C for 48 h, ground and stored at 4 °C. Control fruits were subjected to the same treatment without being inoculated.

Toxin analyses

Water for HPLC mobile phase was purified in a Milli-Q system (Millipore, Bedford, MA). Organic solvents (HPLC grade) were purchased from Merck (Switzerland). The standards for beauvericin (catalogue no. B7510) and enniatin mixture (catalogue no. E3643) were purchased from Sigma Chemical Co. (St. Louis, MO). The enniatin mixture contained enniatin B – 19%, enniatin B_1 – 54%, enniatin A – 3%, and enniatin A_1 – 20%.

To extract BEA and enniatin, 5 g of each sample (of both melon and fungal cultures on rice) was ground and homogenized in a Ultraturax for 3 min with 25 ml of methanol (99.5%). Samples were filtered through Whatman no. 4 filter paper and methanol was removed under reduced pressure. Extracts were resuspended in 3 ml of methanol and pre-purified once on a C18 column (Varian Inc., Palo Alto, CA). The column was activated with 3 ml of methanol and the loaded extract eluted with 2 ml of methanol. The extract was concentrated to 1 ml and filtered through an Acrodisk filter (pore size, 0.22 μm) (Millipore, Jonezawa, Japan) before high-performance liquid chromatography (HPLC). Twenty microlitres of the extract were loaded onto the column. Beauvericin and enniatin (B, B_1, and A_1) analyses were performed according to Monti et al. (2000) with minor modifications. HPLC analyses were performed using LC-10AD pumps and a diode array detector (DAD) from Shimadzu (Japan). A Shiseido Capcell Pak C_{18} ($250 \times 4.6\ mm^2$, 5 μm) column was used. HPLC conditions included a constant flow at 1.5 ml min^{-1} and acetonitrile-water (65 : 35 v/v) as a starting eluent system. The starting ratio was kept constant for 5 min and then linearly modified to 70% acetonitrile in 10 min. After 1 min at 70% acetonitrile, the mobile phase was returned to the starting conditions in 4 min. Beauvericin and the enniatins were detected at 205 nm. Toxins were identified by comparing retention times and UV spectra of samples with authentic standards. Further confirmation was obtained by co-injecting standards with each sample. Toxins were quantified by comparing peak areas from samples to a calibration curve of standards. Chemical structures were confirmed with a liquid chromatography-mass spectrometer (LC-MS). HPLC conditions described above for the analytical separations were applied, but a different HPLC system was used. A Perkin–Elmer LC series 200 connected to a 785A UV/VIS detector was coupled with an API-100 single quadruple mass spectrometer (Perkin–Elmer Sciex Instruments, Ontario, Canada). A flow rate of 20 μl min^{-1} was split from the LC eluent into the ion spray source. A probe voltage of 5300 V and de-clustering potential of 50 V were used. Full-scan spectra were acquired from 400 to 800 amu using a step size of 0.5 amu and a dwell time of 4.2 ms. The instrument mass-to-charge ratio scale was calibrated with the ions of the ammonium adducts of polypropylene glycol. The detection limit was 20, 1.3, 3.6, and 1.2 ng g^{-1} for beauvericin, enniatin A_1, enniatin B_1, and enniatin B, respectively.

Results

Data on race classification and toxin production are reported in Table 1. The pathogenicity assays, performed with thirty-nine strains, showed that all four races of the pathogen occurred among this set of isolates. In particular, four strains belonged to race 0, seven strains to race 1, four strains to race 2, and twenty-two to race 1,2; in addition, two non-pathogenic strains were also characterized. For toxin production, among the forty-four strains tested, thirty-six strains produced beauvericin in a range from 1 to 310 $\mu g\ g^{-1}$. In particular, the four strains of race 0 produced beauvericin at 2–115 μg g^{-1}, the seven strains of race 1 produced the toxin from 1 to 100 μg g^{-1}, and all the four strains of race 2 produced beauvericin in a range from 10 to 135 μg g^{-1}. Among the twenty-two strains of race 1,2, only sixteen produced beauvericin, in the range 2–310 μg g^{-1}. In addition, of the two non-pathogenic strains, only one strain produced the toxin (290 μg g^{-1}). Five strains included in this study, isolated from diseased melon plants but not characterized for pathogenicity, produced beauvericin (from 50 to 150 μg g^{-1}). The production of enniatin A_1, enniatin B_1, and enniatin B was also investigated. Enniatin B was the only enniatin produced by these cultures and only by eleven of the cultures with a range of concentration from traces to 60 μg g^{-1} (Table 1).

Melon fruits inoculated with strain ITEM 3464 resulted in contamination with beauvericin. Cantalupo fruits contained 11.2 μg g^{-1} of the toxin (mean value), and Amarillo fruits contained 73.8 μg of beauvericin per gram.

Discussion

This is the first report on beauvericin production by strains belonging to a *forma specialis* of *F. oxysporum*. The data showed that no relationship was found between a specific race and the production of beauvericin, since isolates belonging to all four races were able to produce the toxin. These findings suggest that this metabolite is not related to specificity of the races of *F. oxysporum f. sp. melonis* and the level of production of beauvericin (Table 1) *in vitro* does not appear related to pathogenicity, since the two non-pathogenic isolates gave opposite results; ITEM 3459, can be considered as a high producer, while ITEM 3296 did not produce any detectable amount of toxin. On the other hand, Sagakuchi et al. (2000) reported that beauvericin was highly toxic to melon protoplasts compared with other toxins with known phytotoxic effects, such as fusaric acid and fumonisin B_1 (Desjardins and Hohn, 1997). Moreover, Lemmens et al. (2000), who investigated the mode of action of beauvericin towards biological membranes using artificial membranes as a model, reported that beauvericin is a very potent channel-forming molecule as it induces pores (which are selective for cations and non-permeable anions) even when present in trace amounts. It was concluded that the non-selective toxic activity of beauvericin could be explained by its ability to induce pores in biological membranes, resulting in a disturbance of the normal gradients of physiologically important monovalent cations across membranes (Lemmens et al., 2000). This mechanism could be expressed by beauvericin in melon plants, once the metabolite has passed through the cell wall of the melon and is in contact with the cell membrane. Such a process could determine a role for beauvericin in the pathogenicity caused by strains of *F. oxysporum f. sp. melonis* that produce the toxin.

Some of the strains also produced enniatin B, a toxin that has been shown to be phytotoxic (Burmeister and Plattner, 1987; Gaumann et al., 1960) and to have a specific role in plant pathogenesis (Herrmann et al., 1996). Although these depsipeptides do not seem to be related to the ability of pathogens to cause disease since they are produced by all races of the pathogen (Table 1), the ability of some strains to produce both the toxins could provide them with further tools for causing phytotoxic effects in melon plants. In addition, the occurrence of beauvericin in the melon fruits artificially inoculated by strain ITEM 3464 showed that the fungus could accumulate the toxin in the fruits. This is important, since melon fruits are especially sensitive to fungal attack during all stages from harvest to storage, as a result of mechanical damage during processing, which helps the fungus to penetrate the fruits through wounds around the peduncle (Brigati and Gori, 1986).

The ability of *F. oxysporum f. sp. melonis* strains to produce both beauvericin and enniatin B, besides confirming the importance of beauvericin as a common metabolite of phytopathogenic *Fusarium* species, suggests that further investigations on the possible additive or synergistic effects of both the toxins on melon plants should be carried out.

Acknowledgments

This work was supported by COST-835 "Agriculturally Important Toxigenic Fungi" and by EU project (QLK1-CT-1999-01380).

References

Abdalla MY, Al-Rokibah A, Moretti A and Mulè G (2000) Pathogenicity of toxigenic *Fusarium proliferatum* from date palm in Saudi Arabia. Plant Disease: 84: 321–324

Appel DJ and Gordon TR (1996) Relationships among pathogenic and nonpathogenic isolates of *Fusarium oxysporum* based on the partial sequence of the intergenic spacer region of the ribosomal DNA Molecular Plant–Microbe Interaction 9: 125–138

Bernardini M, Carilli A, Pacioni G and Santurbano B (1975) Isolation of beauvericin from *Paecilomyces fumosoroseus*. Phytochemistry 14: 1865

Bouhot D (1981) Some aspects of the pathogenic potential in *formae speciales* and races of *Fusarium oxysporum* on *Cucurbitaceae*. In: Nelson PE, Toussoun TA, and Cook RJ (eds) *Fusarium*: Diseases, Biology and Taxonomy (pp 318–326) Pennsylvania State University, University Park, PA

Brigati S and Gori P (1986) Prevenzione dei marciumi da*Fusarium oxysporum* dei meloni dopo la raccolta. Informatore Fitopatologico 5: 33–34

Burmeister HR and Plattner RD (1987) Enniatin production by *Fusarium tricinctum* and its effects on germinating wheat seeds. Phytopathology 77: 1483–1487

Desjardins AE and Hohn TM (1997) Mycotoxins in plant pathogenesis. Molecular Plant–Microbe Interaction 10: 147–152

Gaumann E, Naef-Roth S and Kern H (1960) Zur phytotoxischen Wirksamkeit der Enniatine. Phytopathologische Zeitschrift 40: 45–51

Gupta S, Krasnoff SB, Underwood NL, Renwick JAA and Roberts DW (1991) Isolation of beauvericin as an insect toxin from *Fusarium semitectum* and *Fusarium moniliforme* var. *subglutinans*. Mycopathologia 115: 185–189

Hamill RL, Higgens CE, Boaz HE and Gorman M (1969) The structure of beauvericin: A new depsipeptide antibiotic toxic to *Artemia salina*. Tetrahedron Letters 49: 4255–4258

Herrmann M, Haese A and Zocher R (1996) Effect of disruption of the enniatin synthetase gene on the virulence of *Fusarium avenaceum*. Molecular Plant–Microbe Interaction 9: 226–232

Jacobson DJ and Gordon TR (1991) *Fusarium oxysporum f. sp. melonis*: A case study of diversity within a forma specialis. Phytopathology 81: 1064–1067

Jacobson DJ and Gordon TR (1990) Further investigations of vegetative compatibility within *Fusarium oxysporum f. sp. melonis*. Canadian Journal of Botany 68: 1245–1248

Lemmens M, Kouri K and Lemmens-Gruber R (2000) Patch clamp studies on the electrophysiological properties of the beauvericin channel-incorporated in artificial biological membranes. 6th European *Fusarium* Seminar, 11–16 September 2000 (pp 43–44) Berlin (Abstract)

Logrieco A, Moretti A, Castella G, Kostecki M, Golinski P, Ritieni A and Chelkoswski J (1998) Beauvericin production by *Fusarium* species. Applied and Environmental Microbiology 64: 3084–3088

Logrieco, A., Moretti A, Altomare C, Bottalico A and Carbonell Torres E (1993) Occurrence and toxicity of *Fusarium subglutinans* from Peruvian maize. Mycopathologia 122: 185–190

Monti SM, Fogliano V, Logrieco A, Ferracane R and Ritieni A (2000) Simultaneous determination of beauvericin, enniatins, and fusaproliferin by high performance liquid chromatography. Journal of Agricultural and Food Chemistry 48: 3317–3320

Moretti A, Logrieco, A, Bottalico, A, Ritieni A, Fogliano V and Randazzo G (1996) Diversity in beauvericin and fusaproliferin production by different populations of *Gibberella fujikuroi* (*Fusarium* section *Liseola*). Sydowia 48: 45–56

Moretti, A., Logrieco A, Bottalico A, Ritieni A, Randazzo G and Corda P (1995) Beauvericin production by *Fusarium subglutinans* from different geographical areas. Mycological Research 99: 282–286

Nelson PE, Toussoun A and Marasas W (1983) *Fusarium* species: An illustrated manual for identification. The Pennsylvania State University Press, University Park, PA

Plattner RD and Nelson PE (1994) Production of beauvericin by a strain of *F. proliferatum* isolated from corn fodder for swine. Applied and Environmental Microbiology 60: 3894–3896

Risser G, Banihashemi Z and DavisDW (1976) A proposed nomenclature of *Fusarium oxysporum f.sp. melonis* races and resistance genes in *Cucumis melo*. Phytopathology 66: 1105–1106

Sagakuchi M, Moretti A, Endo E, Matsuda Y, Toyoda H and Ouchi S (2000) An approach to the use of plant sensitivity for simple detection of mycotoxins. In: Proceedings of First Asian Conference of Plant Pathology, August 2000 (pp 262–279), Kuala Lumpur, Malaysia

European Journal of Plant Pathology **108:** 667–673, 2002.

Epidemiology of *Fusarium* infection and deoxynivalenol content in winter wheat in the Rhineland, Germany

B. Birzele[1], A. Meier[2], H. Hindorf[2], J. Krämer[1] and H.-W. Dehne[2]
[1]*Department of Agricultural and Food Microbiology, Institute for Plant Diseases, University of Bonn, Meckenheimer Allee 168, 53115 Bonn, Germany (Phone: +49 228 737729/2110; Fax: +49 228 739592; E-mail: b.birzele@uni-bonn.de);* [2]*Department of Plant Diseases, Institute for Plant Diseases, University of Bonn, Nussallee 9, 53115 Bonn, Germany*

Key words: Fusarium, epidemiology, inoculum sources, deoxynivalenol

Abstract

Details of our long-term research programme concerning the epidemiology of *Fusarium* spp. and mycotoxin production are summarized. Evaluation of the occurrence of *Fusarium* spp., mainly on winter wheat (*Triticum aestivum*), was carried out by investigating *Fusarium* infection and mycotoxin contamination. Two to 15% of grains were infested during 1995–1998 at three climatologically differing localities of the Rhineland, Germany. Disease progress was accelerated by rainfall during the flowering season. The species most frequently isolated were *Fusarium avenaceum*, *F. poae*, *F. culmorum* and *F. graminearum*. The mean deoxynivalenol (DON) content varied from 19 $\mu g\,kg^{-1}$ (1995) to 310 $\mu g\,kg^{-1}$ (1998) and was not always correlated with disease severity. Organic farming systems showed lower rates of infection with ear blight and lower mycotoxin contamination than conventional farming systems.

Introduction

Since the last decade, research work has been conducted on epidemiological aspects of ear blight in wheat and barley. Infection methods under controlled conditions were carried out to study the effect of different inoculum sources (Engels and Krämer, 1996), the climatological environments for disease intensity and the distribution of the fungi in plants during the different growth stages (Meier et al., 2000). Furthermore, different cultivars were studied concerning their susceptibilities to *Fusarium* infection (Stähle et al., 1998; Lienemann et al., 2000). Artificial infection experiments on barley with *Fusarium culmorum* and *F. avenaceum* indicated that infection rates on grains were highest when the inflorescence was inoculated, while lower rates were obtained when seeds and soil were infected during sowing time (Hindorf, 1995). Identification of different genera and species occurring on tissues of the wheat crop has been extended from morphological and biochemical characteristics to immunological characteristics by means of ELISA (Schwabe et al., 1993a) and latex agglutination tests (Schwabe et al., 1992; 1993b; 1994). Furthermore, molecular characteristics were used for identification with PCR (Schütze et al., 1997; Muthomi et al., 2000; Birzele et al., 2000a,b). The increased demand for data on mycotoxin contamination of grains led to the application of different methods of mycotoxin quantification, e.g. ELISA and chromatographic methods (Schwabe and Krämer, 1995; Berleth et al., 1998; Meier et al., 1999; Birzele et al., 2000a). The impact of suboptimal storage conditions on mycotoxin contamination and *Fusarium* biomass has also been investigated (Berleth et al., 1998; Backes and Krämer, 1999; Birzele et al., 1999, 2000a).

Methods and field sites

Investigations of the *Fusarium* mycoflora were carried out during the period 1995–1998 at three localities of the Rhineland, Germany. To compare different farming systems, field experiments with organic and conventional farming systems at two neighbouring sites at Hennef, Sieg (altitude: 65 m, rainfall: 700–750 mm, mean temperature: 9.5 °C) were established. These sites differed in agricultural practices, fertilizers and pesticide input, but not in environmental conditions such as climate and soil. The site at Velbert, Bergisches Land (altitude: 240 m, rainfall: 1200 mm, mean temperature: 9.3 °C) is situated in a wet area and was under an organic farming system. The third site at Blankenheim, Eifel mountains (altitude: 500 m, rainfall: 860 mm, mean temperature: 7.3 °C) with marginal conditions for wheat growing was also chosen for investigations on organic farming.

Plant material and grains were taken for isolating *Fusarium* spp. Shrivelled and broken grains were investigated separately. The infection rate of grains, certified seeds and plant material was determined by incubating 200 grains/certified seeds/plant material pieces per cultivar on selective media. In order to determine the inoculum potential in the soil, organic particles of the soil were fractionated by size (Häni, 1979), surface disinfected and also incubated on selective media (Abildgren et al., 1987; Nirenberg, 1976). *Fusarium* species were differentiated microscopically according to Nelson et al. (1983). The deoxynivalenol (DON) content was quantified by a competitive ELISA, HPLC and also liquid chromatography with tandem mass spectroscopy (LCMSMS).

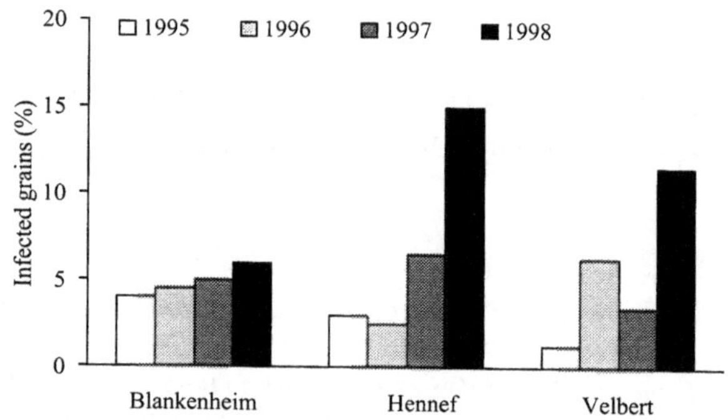

Figure 1. Occurrence of *Fusarium* spp. on winter wheat grains from 1995 to 1998 at three organic farming sites (Blankenheim, Hennef, Velbert).

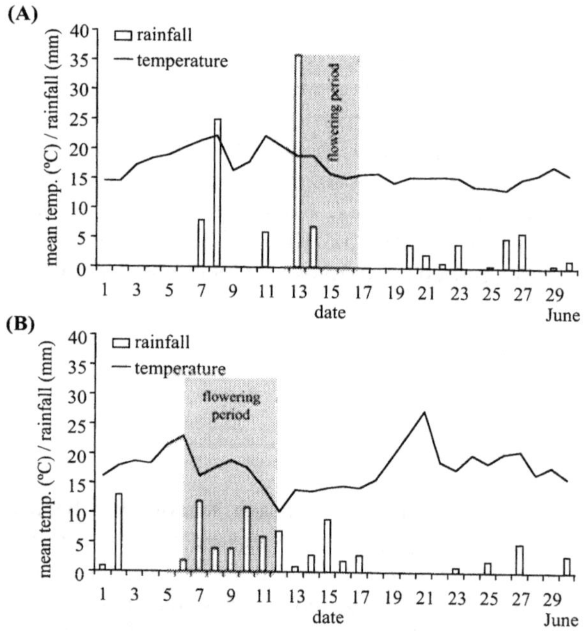

Figure 2. Climatic conditions during the flowering period of wheat at Hennef; (A) 1997, (B) 1998.

Results

The occurrence of *Fusarium* spp. in wheat ears depended mostly on the climatic conditions during flowering (Figures 1 and 2). Figure 2 shows rainfall and temperature at Hennef in June 1997 and 1998. In 1998, continuous rainfall during and after flowering, with lower temperatures and a longer period of open inflorescences, led to grain infection levels of 15% and an average DON content of 310 $\mu g\,kg^{-1}$ at Hennef (Figure 3). Compared to 1998, the days during and after flowering remained dry in 1997, apart from two instances of rain on the first two days of flowering (Figure 2). The average infection with *Fusarium* species in 1997 was lower (3.5–6.5%) and so was the mean DON content (115 $\mu g\,kg^{-1}$). For all three organic farming localities, infection rates were 6.5–15% during 1998 (Figure 1). In the very dry year of 1995, infection rates did not exceed 2–4% of the total harvested crop (Schade-Schütze, 1999) and the mean DON content was 19 $\mu g\,kg^{-1}$ (Backes and Krämer, 1999).

The comparison of infection rates of *Fusarium* species and DON contents showed that they were not always correlated. In 1998, for example, the content of DON was highest at Velbert, with a mean of 395 $\mu g\,kg^{-1}$ compared to 310 $\mu g\,kg^{-1}$ at Hennef. However, greatest *Fusarium* infection in that year was observed at Hennef, with 15% compared to 11.5% at Velbert (Figure 3). At Blankenheim, where conditions for wheat growing are marginal, *Fusarium* infection and DON contents were lowest in 1998. This could be linked to the dry period during flowering, but also to comparatively low temperatures leading to low disease severity of *Fusarium* species and also other plant pathogens.

The different *Fusarium* species were scored and identified at all localities. The distribution of *Fusarium* species for example at Hennef, Rhineland, is shown in Figure 4. Out of the total *Fusarium* spp. infection rate of 6.5% (1997) and 15% (1998), *F. avenaceum* occured in both years at a frequency of 30% (1997) and 51% (1998). The percentage of both *F. culmorum* and *F. graminearum* was similar with 35% in 1997 and 37% in 1998. However, a shift of the population from *F. culmorum* to a higher percentage of *F. graminearum* was observed.

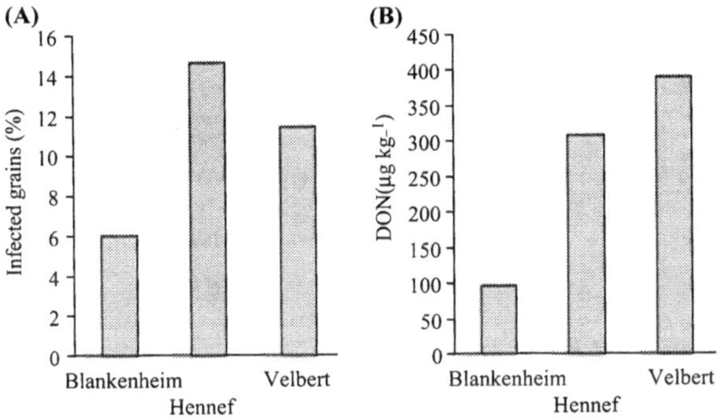

Figure 3. Comparison of (A) *Fusarium* grain-infection and (B) DON content of wheat grains in 1998 in organic farming systems at three localities (Blankenheim, Hennef, Velbert).

The low occurrence of plant pathogens and the dry weather conditions during flowering influenced the yield at Blankenheim in 1998, which was highest with 60 dt ha^{-1} in that specific year. Table 1 shows the yield figures of the organic farming system from 1995 to 1998: no conclusions on the influence of *Fusarium* ear blight on the yield could be made. Besides other parameters, rain, temperature and soil composition at Hennef were optimal for wheat growing, and therefore the mean yield was highest. At Velbert, low yields varying from 30 to 46 dt ha^{-1} with a mean of 41 dt ha^{-1}, were constantly achieved probably because of the high rainfalls. Yields at Blankenheim varied from 26 to 62 dt ha^{-1} showing high uncertainty for stable yields.

Investigations on certified seeds, soil, stem-bases, leaves of different growth stages, as well as grains, for the presence of *Fusarium* spp. (Figure 5) showed that *F. avenaceum* was detected in all samples, but to higher amounts in soil, leaves and grains. *F. graminearum* and *F. culmorum* were observed in soil, leaves and grains, but not in the certified seeds. This indicates that the infection of grains with *Fusarium* species primarily

Table 1. Average wheat yield at three localities (Blankenheim, Hennef, Velbert) in organic farming systems from 1995 to 1998

Localities	Yield (dt ha^{-1})				
	1995	1996	1997	1998	Average
Blankenheim	42	62	26	60	48
Hennef	58	75	61	57	63
Velbert	43	46	44	30	41

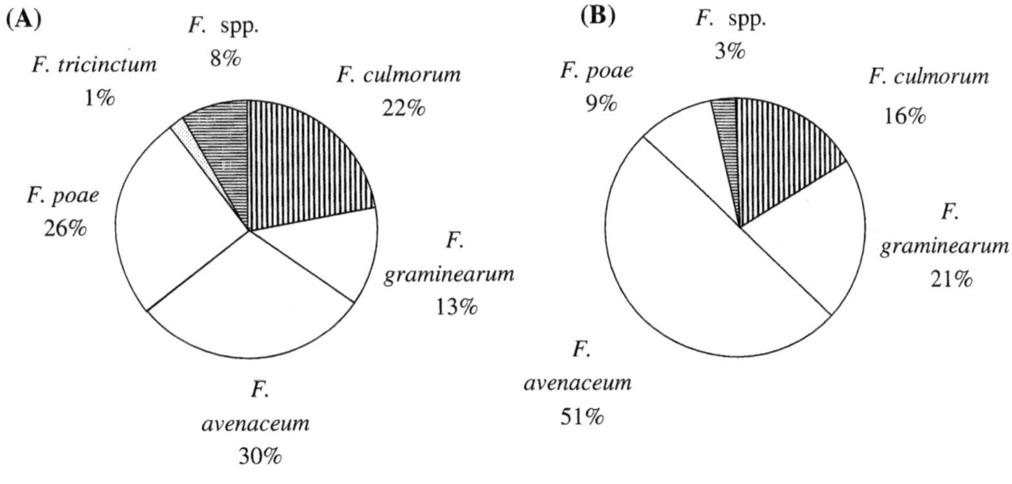

Figure 4. Occurrence of *Fusarium* spp. on wheat grains at Hennef (A) in 1997 and (B) in 1998.

originated from the soil ($r = 0.74$; $P \leq 0.01$), but not from certified seeds ($r = 0.24$; $P \leq 0.01$), from which wheat plants descended. Unlike *Fusarium* spp., *M. nivale* was not isolated from soil. Infection rates for *M. nivale* were highest at the stem base (89%), followed by grains (38%), leaves of GS 75-85 (26%) and certified seeds (6%). *M. nivale* infection on grains and certified seeds correlated ($r = 0.88$; $P \leq 0.01$) confirming that infection with *M. nivale* starts on the certified seeds.

Comparing the occurrence of *Fusarium* and the content of DON in organic and conventional farming systems at Hennef (Figure 6), these were higher in the conventional system, where no specific fungicide was used to control ear blight in the years 1997 and 1998. Application of fungicides during flowering, e.g. tebuconazole, reduced the disease intensity of ear blight to 2/3 of the untreated fields. The organic farming field site showed a much lower occurrence of *Fusarium* and DON content than the conventional farming sites.

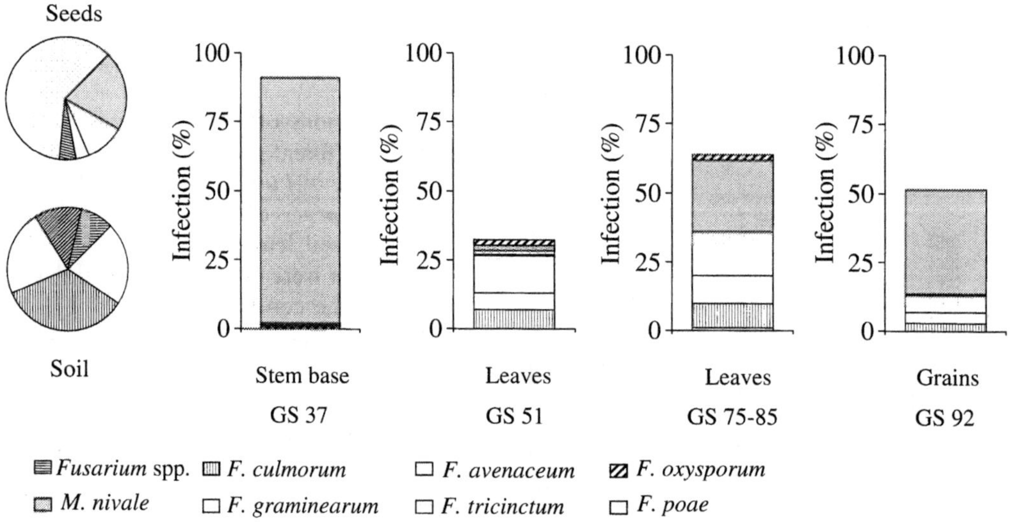

Figure 5. *Fusarium* species identified from soil, certified seeds, plant tissues and grains of wheat at Hennef in 1998.

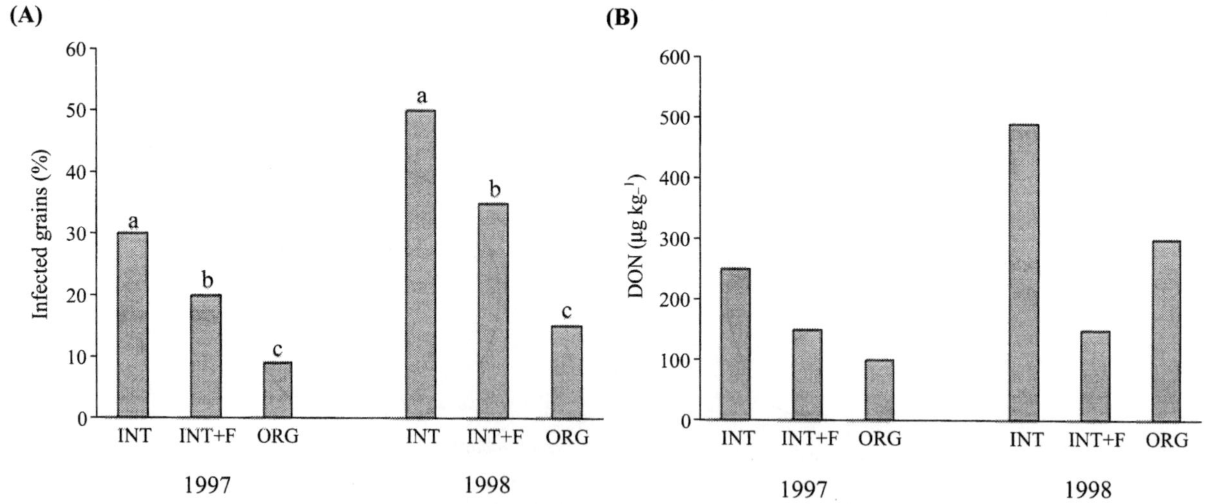

Figure 6. Occurrence of (A) *Fusarium* spp. on wheat grains and (B) DON content of grains in an organic and a conventional farming system at Hennef in 1997 and 1998 (INT = conventional without fungicide control, INT + F = conventional with fungicide control, ORG = organic farming); Tukey test, $P \leq 0.05$.

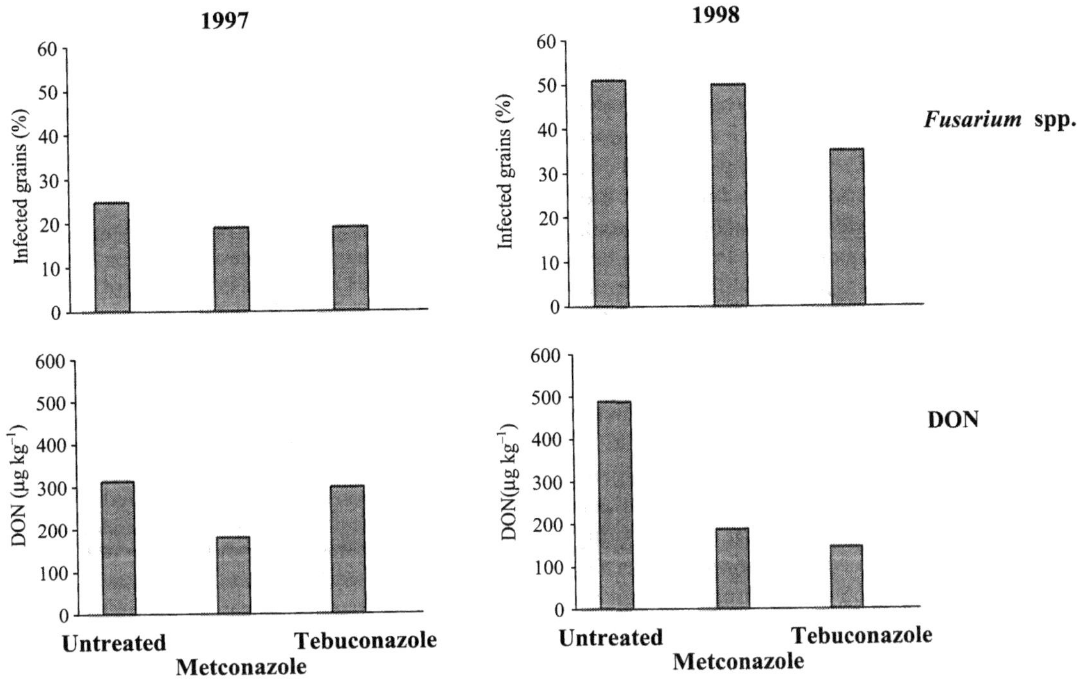

Figure 7. Effect of fungicides on the infection of wheat grains with *Fusarium* spp. and the grain DON content at Hennef in 1997 and 1998.

However, in 1998, when *Fusarium* infection of grains reached 50% in the conventional farming site, fungicide application led to lower DON contents compared to the organic farming site.

In conventional farming systems, the ability of modern fungicides to control ear blight was assessed. During the two years of experiments, metconazole and tebuconazole were applied at the recommended doses at flowering. A certain success was achieved in reducing ear infection at harvesting time (Figure 7). In 1998, high overall infection rates of wheat grains with *Fusarium* spp. were reduced by the application of tebuconazole. DON content of the grains was reduced with both metconazole and tebuconazole. Interestingly, with the lower grain infection levels in 1997 the effectiveness of the fungicides for reducing disease and reducting DON content was lower.

Discussion

At the organic farming sites, *Fusarium* infection of grains varied between 2% and 15% during 1995 and 1998. The occurrence of *Fusarium* spp. in the ear depended mostly on climatic conditions during flowering. In 1998, at Hennef, constant rainfall during and after flowering resulted in an average infection rate of grains of 15% and a mean DON content of 310 µg kg^{-1}. Conclusions on the influence of *Fusarium* ear blight on yield could not be made at infection levels of grains between 2% and 15%. Whereas no correlation was found between the incidence of *Fusarium* spp. on certified seeds and grains ($r = 0.24$; $P \leq 0.01$), there was a strong correlation between the incidence of *M. nivale* on certified seeds and grains ($r = 0.88$; $P \leq 0.01$). A correlation was also determined for the incidence of *Fusarium* spp. on crop residues in the soil (organic particles) and the grains ($r = 0.74$; $P \leq 0.01$). This indicates that the infection of grains mainly originates from the soil, while infection with *M. nivale* originates from the certified seeds.

Comparing the organic farming with the conventional farming system, *Fusarium* infection of grains was significantly higher in the conventional farming system irrespective of fungicide treatment. In years such as 1998, with extremely high infection levels of grains (50%) in the conventional farming system, fungicide treatment reduced the DON levels in the grain below those from the organic farming site. In 1997, when average infection rates of grains occurred, DON contents remained lowest in the organic farming system. Further research on the influence of different fungicides on *Fusarium* infection and DON contents is needed.

Our long-term research work on the occurrence of *Fusarium* spp. on wheat and their mycotoxin production will continue. Further work will study regional effects on the intensity of *Fusarium* spp. Methods will be improved for the detection of the fungi and their mycotoxin production. Besides microscopy, molecular methods such as PCR and quantitative PCR, have been developed for the precise diagnosis of different species. Other points of interest are the epidemiology of the infection process during flowering and the penetration of the fungus into the plant. Furthermore, sources of inoculum and their distribution as well as the influence of agricultural practices are of particular interest. As a result of the field experiments carried out in organic and conventional farming systems in the Rhineland, the influences of crop rotation, cultivar susceptibility, soil preparation (tillage), weed populations as alternate hosts for *Fusarium* spp., the use of fungicides on the infection rate of *Fusarium* spp. and contamination with DON have been investigated.

The influence of sub-optimal storage conditions on the infection rate of *Fusarium* spp., *Penicillium* and *Aspergillus* spp. will be further investigated and common mycotoxins such as DON, nivalenol (NIV), T-2 toxin and ochratoxin A (OTA), produced by *Penicillium* and *Aspergillus* species, will be analysed in relation to fungal biomass. Furthermore, investigations on the influence of fungi, mycotoxins and sub-optimal storage of wheat on the baking quality and rheological properties of contaminated flours will be continued as part of our research on food quality and food safety.

Acknowledgements

Most of the studies were supported by the Deutsche Forschungsgemeinschaft (DFG) within the DFG research group OSIOL ('Optimizing strategies in organic farming') and by the Ministerium für Umwelt, Raumordnung und Landwirtschaft (MURL) of NorthrhineWestphalia (grants to H.-W. Dehne and J. Krämer).

References

Abildgren MP, Lund F, Thrane U and Elmholt S (1987) Czapek-Dox agar containing iprodione and dichloran as a selective medium for the isolation of Fusarium species. Letters in Applied Microbiology 5: 83–86

Backes F and Krämer J (1999) Mikrobiologische und mykotoxikologische Qualität von Winterweizen aus Organischem Landbau als Rohstoff für Lebensmittel. Getreide Mehl und Brot 53: 197–201

Berleth M, Backes F and Krämer J (1998) Schimmelpilzspektrum und Mykotoxine (Deoxynivalenol und Ochratoxin A) in Getreideproben aus ökologischem und integriertem Anbau. Agribiological Research 51: 369–376

Birzele B, Meier A and Krämer J (1999) Einfluss suboptimaler Lagerung auf die Entwicklung von DON in Weizen aus Organischem Landbau. In: Proceedings of the 21st Mykotoxin-Workshop (pp 26–29) Jena, Germany

Birzele B, Prange A and Krämer J (2000a) Deoxynivalenol and ochratoxin A in German wheat and changes of level in relation to storage parameters. Food Additives and Contaminants 17: 1027–1035

Birzele B, Prange A, Schönling J and Krämer J (2000b) Development of deoxynivalenol contents in relation to the PCR detection of potentially trichothecene producing *Fusarium* spp. during storage of wheat. Mycotoxin Research 16A: 46–49

Engels R and Krämer J (1996) Incidence of Fusaria and occurrence of selected *Fusarium* mycotoxins on *Lolium* spp. in Germany. Mycotoxin Research 12: 31–40

Häni F (1979) Über Getreidefusariosen in der Schweiz: Saatgutbefall, Ährenbefall und Bodenkontamination. Zeitschrift für Pflanzenkrankheiten und Pflanzenschutz 87: 257–280

Hindorf H (1995) Ear infection of wheat and barley with *Fusarium* spp. in the Rhineland/Germany. In: Proceedings of the International Seminar on *Fusarium* Mycotoxins, Taxonomy and Pathogenicity (p 89) Bari, Italy

Lienemann K, Oerke E-C and Dehne H-W (2000) Effect of differences among wheat genotypes on the occurrence and damage caused by *Fusarium* head blight. In: Proceedings of the 6th European *Fusarium* Seminar (p 26) Berlin, Germany

Meier A, Birzele B, Oerke E-C and Dehne H-W (1999) Auftreten von *Fusarium* spp. und Mykotoxingehalte von Winterweizen und Möglichkeiten zur Bekämpfung. In: Proceedings of the 21st Mykotoxin-Workshop (pp 19–25) Jena, Germany

Meier A, Birzele B, Oerke E-C and Dehne H-W (2000) Impact of growth conditions on the occurrence of *Fusarium* spp. and the mycotoxin content in wheat. Mycotoxin Research 16A: 12–15

Muthomi JW, Schütze A, Dehne H-W, Mutitu EW and Oerke E-C (2000) Characterization of *Fusarium culmorum* isolates by mycotoxin production and aggressiveness to winter wheat. Journal of Plant Diseases and Protection 107: 113–123

Nelson PE, Toussoun TA and Marasas WFO (1983) *Fusarium* species. Illustrated Manual for Identification. Pennsylvania State University Press, University Park London, Great Britain

Nirenberg H (1976) Untersuchungen über die morphologische und biologische Differenzierung in der *Fusarium*-Sektion Liseola. Mitteilungen der Biologischen Bundesanstalt für Land- und Forstwirtschaft 169: 1–117

Schade-Schütze A (1999) Auftreten und biologische Charakterisierung von *Fusarium*-Arten im Weizenanbau. PhD thesis, University of Bonn, Germany

Schütze A, Oerke E-C and Dehne H-W (1997) Isolation and differentiation of *Fusarium* spp. and *Microdochium nivale* on winter wheat in Western Germany. In: Proceedings of the 5th European *Fusarium* Seminar (pp 615–616) Szeged, Ungarn

Schwabe M, Kamphuis H, Trümner U, Offenbächer G and Krämer J (1992) Comparison of the Latex Agglutination Test and the Ergosterol Assay for the detection of moulds in foods and feedstuffs. Food and Agricultural Immunology 4: 19–25

Schwabe M, Menke J and Krämer J (1993a) Einsatz des EPS-ELISA zur Bestimmung von Schimmelpilzen der Gattungen *Aspergillus* und *Penicillium* in Lebensmitteln. Archiv für Lebensmittelhygiene 44: 64–66

Schwabe M, Rath F, Golomb A, Engels R and Krämer J (1993b) Immunologischer Schnellnachweis von *Fusarium* in Getreide. Monatsschrift für Brauwissenschaft 11: 408–415

Schwabe M, Fenz R, Engels R, Krämer J and Rath F (1994) Nachweis von *Fusarium* auf Braugerste mit dem EPS-Latex-Agglutinations-Test. Monatsschrift für Brauwissenschaft 5: 160–164

Schwabe M and Krämer J (1995) Influence of water activity on the production of T-2 Toxin by *Fusarium sporotrichioides*. Mycotoxin Research 11: 35–39

Stähle A, Birzele B, Schütze A, Oerke E-C, Krämer J and Dehne H-W (1998) Optimierungsstrategien im Organischen Landbau: Auftreten von *Fusarium* spp. und DON-Konzentrationen in Winterweizen 1997. In: Proceedings of the 20th Mykotoxin-Workshop (pp 262–266) Detmold, Germany

European Journal of Plant Pathology **108:** 675–684, 2002.

Role of deoxynivalenol in aggressiveness of *Fusarium graminearum* and *F. culmorum* and in resistance to *Fusarium* head blight

Ákos Mesterházy
Cereal Research non-profit Company, P.O. Box 391, 6701 Szeged, Hungary (Fax: +3662420101; E-mail: akos.mesterhazy@gk-szeged.hu)

Key words: mycotoxin, host-pathogen relationship

Abstract

The data available indicate that aggressiveness of *Fusarium graminearum* and *F. culmorum* depends on their deoxynivalenol (DON) and nivalenol-producing capacity: toxin-producing ability correlated closely with the level of aggressiveness measured. This agrees well with other literature findings. However, the resistance of a cultivar influenced DON production significantly. In the most resistant genotypes, toxin contamination remained near zero, whereas the same isolates and inoculum produced very high toxin levels in susceptible cultivars. As toxin levels were correlated with the ratio of *Fusarium*-damaged kernels (FDK) and this ratio is very low in highly resistant cultivars, the conclusion is that the level of resistance level is more important in governing DON accumulation in a given cultivar than is the aggressiveness of an isolate. In susceptible cultivars, DON producing ability is decisive, but in highly resistant cultivars resistance is the major factor in suppressing disease development and DON accumulation. In different years, the same FDK values were associated with different DON concentrations and this depended very much on the precipitation towards the end of May, the time of inoculation.

Introduction

Following the large Fusarium head blight (FHB) epidemics in the USA, Canada, China AND Europe (including Germany, Hungary, Romania and Croatia) significant research has been devoted to the breeding of resistant cultivars. However, alternative solutions could originate from a better understanding of the host–pathogen relationship, the nature of resistance and an understanding of the role of deoxynivalenol (DON) and other trichothecenes in pathogenesis.

The resistance of wheat to FHB is a complex phenomenon. The forms, types or components of physiological resistance (Mesterházy, 1995; 2001) are: (i) resistance to initial infection (Schroeder and Christensen, 1963); (ii) resistance to spreading (Schroeder and Christensen, 1963); (iii) resistance to kernel infection (Mesterházy, 1995; Mesterházy et al., 1999); (iv) tolerance to infection (Mesterházy, 1995; Mesterházy et al., 1999) and (v) resistance to DON accumulation (Miller et al., 1985).

Research on two other forms of resistance is in progress. Resistance to late blighting means low *Fusarium*-damaged kernel (FDK) values even when a long rainy period occurs at harvest (Mesterházy, unpublished data). Such resistance contrasts with cultivars with low FHB values, but high FDK values at harvest. The resistance to head death above the infection point of the head means that transport of assimilates is still possible following infection of the rachis. As a result, seed size is near normal. In susceptible genotypes, the transport vessels cease function, mycelial masses inhibit transport and the grains shrivel even though they are free of infection. However, the relationship between these parameters and their genetic background is not clear. The best-known component is the resistance to pathogen spread; all genetic work refers to this form of resistance. The genetic inter-relations of these components are not known. In medium-susceptible, medium-resistant or susceptible genotypes, resistance components have been identified and provide information about the risks of the given genotypes. In the most resistant cultivars, however, there are no significant yield losses, the ratio of FDK is very low, FHB severity is small and toxin contamination is also very low. It seems that at high resistance the significance of these components is much less or negligible (Mesterházy et al., 1999).

The morphology of a cereal crop also plays a role in natural infection. Dwarf types are more susceptible than tall cultivars and wheat genotypes with awns are more at risk than tall or awnless types (Mesterházy, 1987; 1995). Cultivars with longer internodes below the heads are less infected (Parry et al., 1995). Resistance at the seedling stage, crown rot resistance and resistance of the leaves has also been demonstrated, but useful relationships with FHB resistance for breeding were not found (Mesterházy, 1983; 1987).

Materials and methods

The methods for fieldwork in resistance research, toxin and ergosterol evaluation have been described by Mesterházy et al. (1999), Lamper et al. (2000) and Mesterházy and Bartók (1996, 1997, 2001). Inoculation was made by spraying the heads and covering them with polythene bags for 24 h. Each isolate was used in three replicates. Fusarium head blight was evaluated as a percentage of the spikelets of 15–20 heads 10, 14, 18, 22 and 26 days after inoculation (full flowering). Their mean value was used in the analyses and AUDPC values were also calculated. From each group of heads, ten spikelets were separated randomly and threshed at low wind speed by a Wintersteiger head thresher 'Seedboy'. The ratio of FDK, that is, the ratio of visually infected scabby or tombstone kernels, was measured. The weight of grains was measured to provide an indication of any yield responses. The combined samples of the three replicates were evaluated for DON and ergosterol content with HPLC. For this 6 g grain was used, milled, and 5 g was evaluated.

Results

Instability of DON production

The DON-producing ability of isolates was not consistent. Very large differences occurred under different conditions. One of the main factors was the interrelationship with cultivar resistance. Table 1 shows the mean results of a three-year test for DON with 25 genotypes and four isolates each of *Fusarium graminearum* and *F. culmorum* (Mesterházy et al., 1999). The mean toxin production in the cultivars in response to the same isolates was different, ranging from 0.32 $\mu g\ g^{-1}$ (ppm) in the most resistant genotypes to 42 $\mu g\ g^{-1}$ in the most susceptible genotypes. Striking differences were also found for the mean performance of isolates with different toxin-producing capacities. It seems that the characterization of DON production for an isolate is much more accurate than for a randomly chosen wheat cultivar. Even when correlations between DON data for isolates and genotypes are significant, there remain differences in toxin concentrations between wheat genotypes when values for mean toxin concentrations in isolates are very similar (for example, see data for *F. graminearum* isolates 40 and 44). We do not know the cause of the three-fold difference in DON concentrations in wheat cultivars for isolates from Bence or Csaba, but the conclusion is clear: to understand fully the influence of cultivar on toxin production we need to study more isolates. The same conclusion is valid for isolate aggressiveness where more cultivars are necessary to gain reliable conclusions. This work is usually not done in this way. Usually one isolate or a mixture of isolates are used on several genotypes, or several isolates are tested on one or more cultivars. As a result, the reliability of the data is low. For this reason, data from such tests are less suitable for drawing general conclusions. It should be pointed out that the *F. graminearum* and *F. culmorum* isolates produced very similar resistance rankings, with correlations normally above $r = 0.90$; $P = 0.001$.

Influence of cultivars

The cultivars influence symptom development and other traits (Table 2). The fact that DON concentrations closely correlate with the ratio of FDK and the extent of FHB proves that toxin contamination is a consequence of the amount of disease observed. Low FDK ratios mean low toxin contamination. When we calculate the ratios for DON/FDK or DON/FHB, we see that in resistant genotypes significantly higher values are needed to produce 1 $\mu g\ g^{-1}$ DON than in susceptible cultivars. In the highly resistant group, 2–8% FHB and 2–4% FDK produces 1 $\mu g\ g^{-1}$ DON. In the more susceptible group, 0.82–1.50% FHB or 0.8–1.5% FDK is needed to produce the same amount. It appears, therefore,

Table 1. FHB resistance of 25 wheat genotypes to *Fusarium* isolates, mean data for DON concentrations in μg g⁻1 ppm, 1994–1996

Genotype	Isolates								Mean
	89.4 Fc	12375 Fc	12377 Fg	207 Fg	12551 Fc	44 Fg	40 Fg	223 Fc	
Sgv-NB * MM-Sum3	0.3	0.0	0.4	0.6	0.3	0.8	0.0	0.2	0.3
Arina	0.4	0.8	1.3	0.0	1.0	1.1	4.2	1.6	1.3
Sgv-NB * MM-Sum3	0.3	1.2	1.5	1.0	1.3	1.4	1.8	5.2	1.7
Ringó Star	0.8	1.9	1.9	2.4	3.0	3.0	1.3	2.2	2.1
RSt- MM/NB	2.1	0.0	0.9	5.8	0.6	3.1	1.5	4.1	2.3
HD84.42	2.3	5.3	3.4	4.2	3.9	4.0	11.0	10.6	5.6
81.60-NB/Kö	4.5	5.4	3.0	3.8	5.2	8.2	5.6	12.8	6.1
P4371.88	2.3	6.2	3.4	4.2	6.1	5.9	10.9	10.2	6.2
SK8090	4.2	2.2	8.5	0.5	1.3	14.1	13.0	8.7	6.6
Siouxland	1.1	1.6	6.4	4.0	4.9	9.4	11.9	19.0	7.3
P2118.89	11.4	8.1	7.3	4.4	2.5	10.9	17.3	14.5	9.5
Sum3^2-81.60	2.5	4.1	2.3	10.0	10.4	12.0	18.2	18.2	9.7
Rechsler	11.5	4.2	3.3	14.8	11.3	18.1	22.8	9.1	11.9
Sgv-GT//Pdj2/Uhr	1.7	8.7	3.9	9.7	12.8	9.9	29.8	25.2	12.7
Kincsö	7.1	8.6	8.1	5.0	13.5	15.7	15.6	35.9	13.7
Kende	9.9	5.8	7.9	13.7	19.9	16.1	22.7	19.2	14.4
Bence	8.2	9.6	7.8	21.8	25.0	43.6	14.8	16.0	18.3
Szöke	5.4	8.9	8.6	27.4	25.7	30.3	29.2	16.5	19.0
78.1.4	11.5	3.8	25.9	7.5	10.7	12.2	39.0	41.7	19.0
Góbé	14.3	11.2	25.4	6.2	18.5	22.9	16.9	39.1	19.3
Jbj-50	9.0	16.2	18.2	19.0	31.8	41.9	19.8	33.6	23.7
Csaba	12.3	17.6	16.3	21.2	38.4	23.2	49.7	67.1	30.7
Zugoly	8.8	19.6	15.6	20.9	35.1	60.3	38.4	59.2	32.2
Örség	9.3	20.5	16.1	30.9	47.1	56.2	53.2	54.6	36.0
Zombor	17.7	18.3	24.1	60.6	52.7	58.1	65.5	41.1	42.3
Mean	6.4	7.6	8.9	12.0	15.3	19.3	20.6	22.6	14.1
LSD 5%									8.6

Fc = *F. culmorum*; Fg = *F. graminearum*.

that DON decomposing mechanisms are present in resistant genotypes (see also Miller and Arnison, 1986).

Climatic effects

Another cause of variation in DON production are the differences in weather and other factors, that occur in different years. The same isolate may, in different years, cause highly significant differences in disease development which are associated with differences in DON concentrations. The highly significant year–isolate interactions show this clearly. This means that more accurate data would be obtained for an isolate if inocula were tested over several years. The influence of year on DON contamination has another aspect. When the general means for the disease parameters and DON production are compared on an annual basis, we see that there are ten-fold differences between the amount of FDK or FHB needed to produce 1 $\mu g\,g^{-1}$ DON. What is the cause of this year effect?

We compared the mean values of FHB, FDK and DON, and yield loss for 1990–2000 with the meteorological data. A close correlation was found between the amount of DON produced and the amount of precipitation between May 20 and 31 ($r = 0.6422$; $P = 0.05$). The amount of precipitation for the whole of May showed no significant relationship with DON contamination level. The between year DON–FHB and DON–FDK correlations were between 0.4 and 0.5. Yield loss was strongly influenced by FHB ($r = 0.7651$; $P = 0.01$) and kernel infection ($r = 0.7996$; $P = 0.01$). Interestingly, June precipitation did not show a relationship with DON or with the other traits. However, the large precipitation from 1st to 20th July significantly increased kernel infection ($r = 0.6520$; $P = 0.05$) and yield loss ($r = 0.7688$; $P = 0.01$). The mean temperatures for the three months separately or together did not significantly influence DON levels or the other traits. The significance of the precipitation in the last third of May was supported by the data set from fungicide

Table 2. Mean data for the FHB test with eight Fusarium isolates, 1994–1996

Genotype	Trait				Ratio for	
	DON $\mu g\, g^{-1}$	FHB %	FDK %	Yield loss %	FHB/DON	FDK/DON
Sgv-NB ∗ MM-Sum3	0.32	2.61	0.92	3.68	8.16	2.87
Sgv-NB ∗ MM-Sum3	1.7	3.45	2.03	7.54	2.03	1.19
Arina	1.28	5.28	3.88	7.61	4.12	3.03
RSt- MM/NB	2.26	7.95	5.02	12.9	3.52	2.22
Rsztár	2.06	10.61	7.08	11.75	5.15	3.44
Siouxland	7.28	13.1	25	25.93	1.8	3.43
SK8090	6.55	14.82	15.54	22.65	2.26	2.37
Sum32̂-81.60	9.72	15.07	11.59	24.11	1.55	1.19
P2118.89	9.54	16.92	30.42	19.8	1.77	3.19
P4371.88	6.15	19.09	17.19	27.32	3.1	2.8
HD84.42	5.59	20.58	17.26	17.43	3.68	3.09
81.60-NB/Kö	6.07	20.83	11.97	24.31	3.43	1.97
Bence	18.34	20.91	16.06	23.97	1.14	0.88
Rechsler	11.9	23.33	28.09	29.86	1.96	2.36
Kincsö	13.67	24.1	18.43	28.42	1.76	1.35
Szöke	18.99	26.15	19.49	33.19	1.38	0.97
Kende	14.41	27.9	22.32	42.53	1.94	1.55
Sgv-GT//Pdj2/Uhr	12.71	28.31	15.73	40.81	2.23	1.24
Jbj-5o	23.69	30.78	42.34	44.69	1.3	1.79
Góbé	19.31	31.51	32.09	37.11	1.63	1.66
78.1.4	19.03	31.72	40.1	39.31	1.67	2.11
Örség	42.25	34.52	43.17	48.21	0.82	1.02
Zombor	18.43	35.74	35.43	39.67	1.94	1.92
Zugoly	32.23	38.37	50.43	46.17	1.19	1.56
Csaba	30.72	38.92	26.45	46.89	1.27	0.86
Mean	13.37	21.7	21.48	28.23	2.43	2.00
LSDS 5 %	8.58	0.98	2.7	3.59		

Correlation analysis

	DON $\mu g\, kg^{-1}$	FHB %	FDK %
FHB %	0.8672		
FDK %	0.8199	0.8358	
Yield loss %	0.8693	0.9527	0.8294

All are significant at $P = 0.001$.

tests performed between 1992 and 1999. We can conclude that DON contamination is influenced more by the amount of precipitation after inoculation than by later rainfall. The late July rainfall increased FDK and also DON, but the correlation for DON was not significant. It seems that FDK and FHB influence DON less than precipitation, though these correlations were not significant. This means that epidemics with similar FDK values can result in ten-fold different levels of DON contamination. Such findings will have important consequences for health.

DON and aggressiveness

The next question concerns the relationship between the DON-producing ability of the pathogen and its disease-causing capacity. A close relationship was usually found between the aggressiveness of an isolate and its DON production (Mesterházy et al., 1999). The exception was *F. culmorum* isolate 89.4F. Here, high aggressiveness was associated with low DON production. Further analysis revealed that this isolate was a nivalenol-producing type with lower DON production. When nivalenol and DON were considered together, the toxin production (total trichothecene) corresponded to the aggressiveness we measured (Figure 1). In 1997, the results were very similar (Table 3). However, in that year, the nivalenol-producing isolate (89.4F) was also a good DON producer. It seems that the production of nivalenol or DON can be dependent on ecological parameters.

The data for the whole period 1990–2000 support these conclusions: correlations between DON

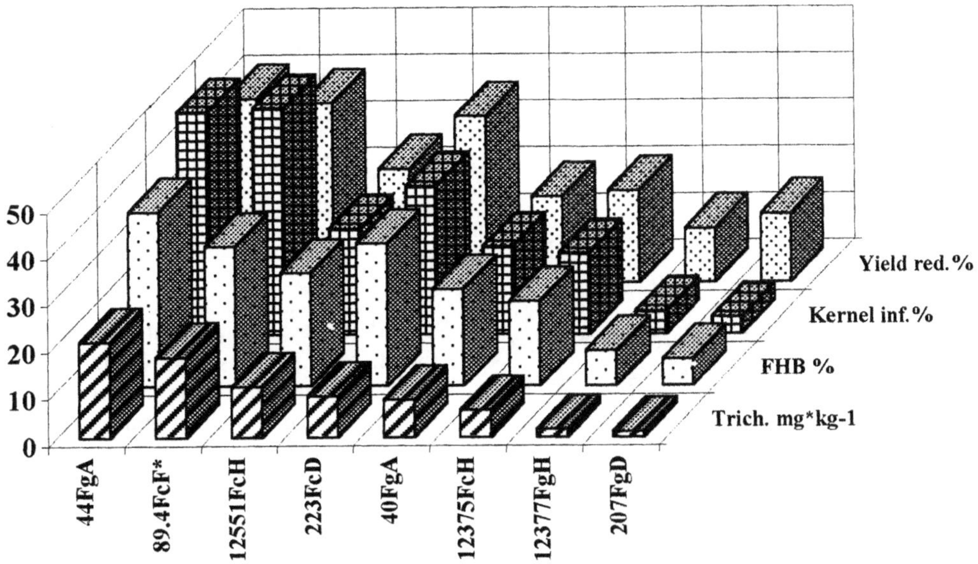

Figure 1. Aggressiveness and trichothecene (DON + NIV) content in FHB tests, 1994.

Table 3. Mean aggressiveness and DON production of *Fusarium* isolates on 22 cultivars, 1997

Isolates	FHB %	FDK %	Yield loss %	DON $\mu g\,g^{-1}$
12551Fc	4.97	5.3	8.51	1.55
39.01Fc	6.89	3.77	3.8	0.84
12377Fg	10.38	22.85	21.26	5.57
223Fg	19.03	35.78	35.24	9.62
40Fg	19.86	38.49	28.31	8.13
89.4FFc	21.52	25.21	31.58	0.16
207Fg	24.48	38.46	39.56	10.73
12375Fc	25.22	37.84	42.88	10.12
Mean	16.54	25.96	26.39	5.84

Correlation analysis

	FHB %	FDK %	Yield loss %
FDK %	0.9075***		
Yield loss %	0.9568***	0.9370***	
DON $\mu g\,kg^{-1}$	0.6635***	0.8478***	0.7591***

*** $P = 0.001$. Fc = *F. culmorum*; Fg = *F. graminearum*.

production and disease development were very close. However, correlations between DON concentration and FDK values were generally closer than between visual symptoms (FHB) and DON. This is to be expected: the same FHB severity may result in very different kernel infection severity and DON concentrations are determined from the grains. Here, resistance to kernel infection plays a crucial role.

Studying the data, we see that there are differences between isolates. Some isolates of *F. culmorum* produce relatively more DON than others and we see differences also in their ability to influence FHB, FDK and yield. Some can cause more FDK, some less FHB and others less yield loss. It seems, therefore, that the *Fusarium* population consists of different individuals or lines having different pathogenic characters including toxin-producing ability. The deviations are not large and the very close correlations show the basic trends that is, higher isolate aggressiveness results in higher FHB severity, higher FDK, higher yield loss and higher DON production.

It seems that ergosterol production (Lamper et al., 2000) is also closely linked with DON and FDK, indicating that aggressiveness is proportional to the fungal biomass in the infected grains (Figure 2). More highly aggressive isolates produced more severe infection, more fungal mass, more ergosterol and more DON.

From 1998, experiments were carried out with other *Fusarium* species. Figures 3 and 4 show the FDK and DON results from the 1998 multi-isolate tests only. Table 4 shows summary data for the 1998 cultivar tests, which indicate a very close relationship between traits. Results show that resistance to *F. graminearum* or *F. culmorum* is linked to resistance to *F. sambucinum, F. sporotrichioides* and *F. verticillioides*. The data for 1999 and 2000 are very similar and confirm that this also applies to *F. poae, F. avenaceum* and *F. nivale*.

Pathogen strains which do not produce DON have low aggressiveness ($r = 0.9448$ between FDK and DON). The small amount of DON associated with infection by these species is due to background infection under epidemic field conditions. This also occurred in other field tests (Bai et al., 2000).

In the fungicide tests, where four *Fusarium* isolates were used, a significant reduction in symptoms was recorded (Table 5). This reduction, measured by FHB, FDK, yield loss and DON contamination, was proportional to the anti-Fusarium activity of the fungicides used. However, Amistar (azoxystrobin) and Kolfugo (carbendazim) produced an increase on susceptible cultivar in DON even though symptom severity was decreased.

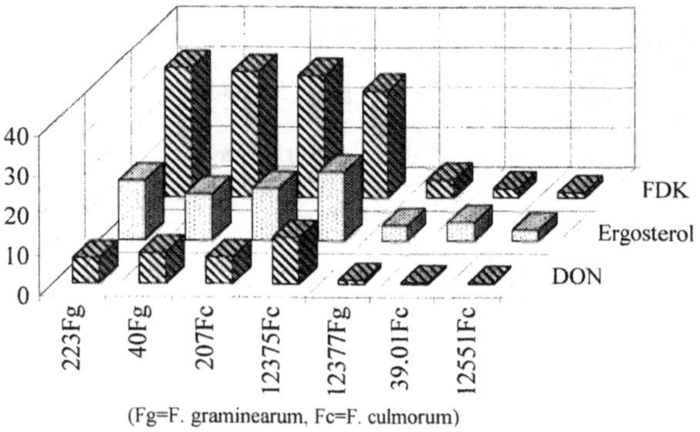

Figure 2. Ergosterol ($\mu g\ g^{-1}$), DON ($\mu g\ g^{-1}$) and FDK (%) values for *Fusarium* isolates, 1998. Fg = *F. graminearum*; Fc = *F. culmorum.*

Discussion

The data describe a complex influence of DON on the disease process. DON and other trichothecenes appear to play an important role in the aggressiveness of *F. graminearum* and *F. culmorum.* DON is a strong protein inhibitor (Snijders, 1994), and this may cause inhibition of enzymatic activity in susceptible hosts, leading to a rapid increase of the disease. This conclusion is supported by Muthomo et al. (2000), who reported a close correlation between aggressiveness and DON production in *F. culmorum* isolates. The results also show that aggressiveness closely correlated

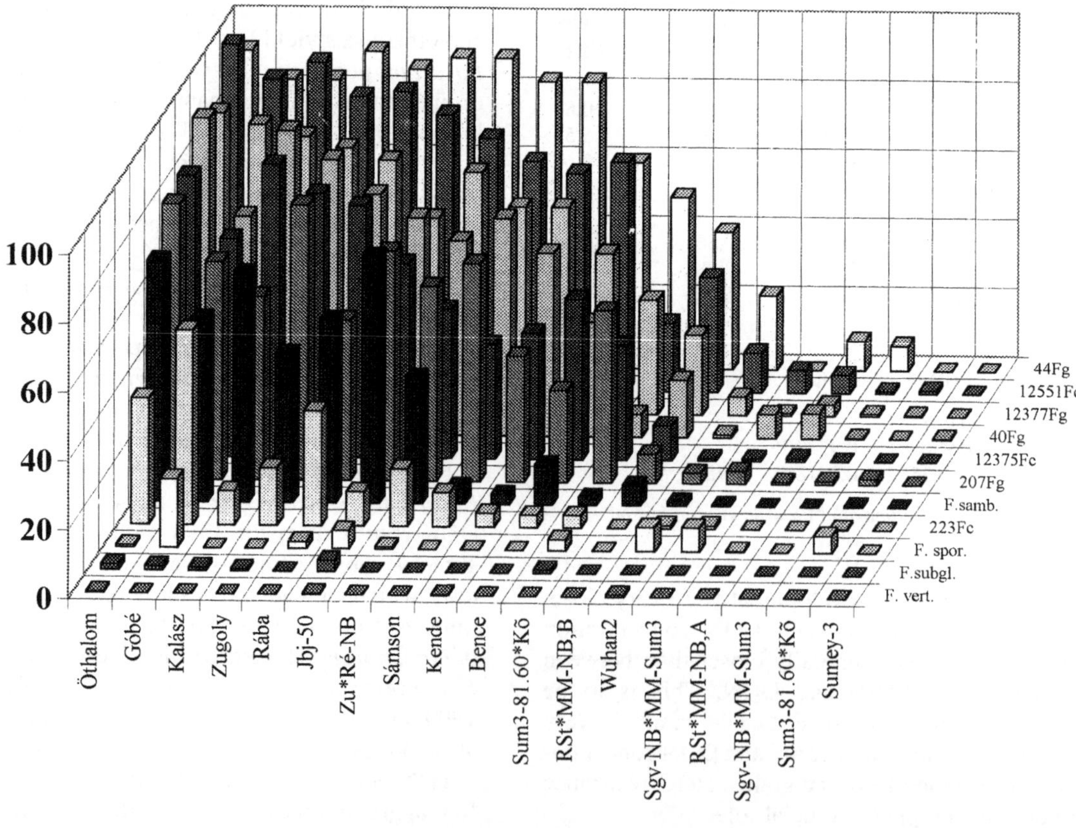

Figure 3. Fusarium head blight resistance in wheat to different *Fusarium* spp., FDK values %, 1998.

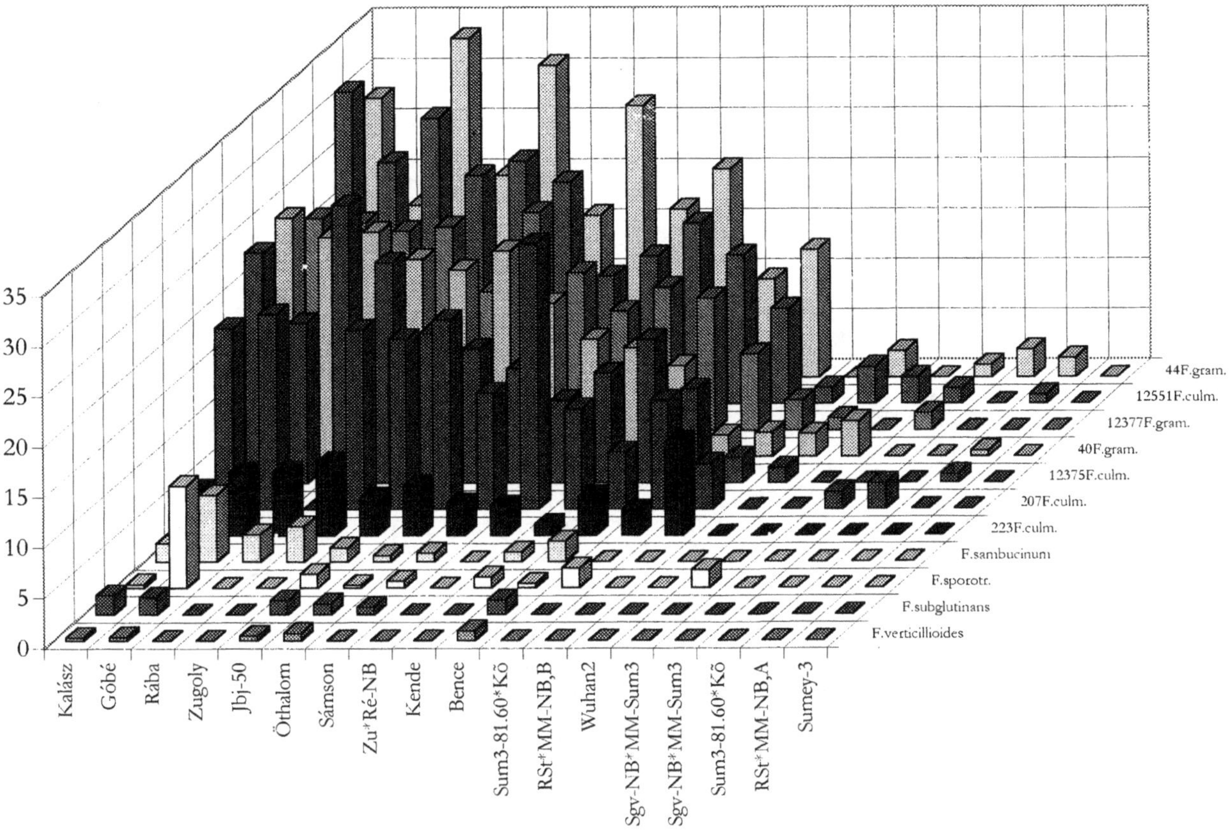

Figure 4. Fusarium head blight resistance in wheat, DON contamination (mg kg^{-1}) 1998.

with DON-producing ability of both species. Similar results were obtained for *F. culmorum* isolates (Gang et al., 1998). The data support the positive role of DON in pathogenesis. We should realize that the different trichothecene toxins have different toxicity in plants and animals. Eudes et al. (1997) reported that DON and 3-AcDON is much more toxic in plants than T-2 toxin or nivalenol.

Proctor et al. (1995) have disrupted the *Tri5* gene which encodes trichodeine synthase and plays a crucial role in the synthesis of the trichothecene skeleton. This resulted in a significant decrease in aggressiveness, again indicating that DON is essential for disease development (Bai et al., 2000; Desjardins et al., 1996; 2000). Their results showed, however, that all aggressiveness was not lost, and the amount of loss was different in different plant species. In wheat, it was more expressed, in other crops less so. These findings support the view that other components of aggressiveness, such as cell wall-degrading enzymes, exist. Alexander et al. (1997) concluded that trichothecenes are not necessary for pathogenicity, but that they increase the extent of the disease. They concluded, however, that breeding for toxin resistance could result in higher resistance to disease. To achieve this, they suggested the incorporation of the *Tr12* transporter gene in plants in the hope that the resulting toxin resistance would cause higher disease resistance and less toxin contamination. As other components also play a role in pathogenesis, the result of the gene transfer might be disappointing. The reduction of the disease could be less than anticipated.

Mirocha et al. (1997) reported that after inoculation, DON was not detected in infected tissue earlier than 48 h. Evans et al. (2000) found a similar time-course after inoculation in barley. Kang and Buchenauer (1999) came to similar conclusions with a detection time of 36 h. Chen et al. (1996) reported 24 h for the first detection of DON after inoculation. The relative late occurrence of DON in disease development suggests that DON does not play a role in the initial phase of infection. Bai et al. (2000) came to the same conclusion for barley. It is important to note that mycelium and

Table 4. Resistance against *Fusarium* species in wheat cultivars to FHB, 1998

Genotype	Traits			
	DON μg g^{-1}	FHB %	Yield loss %	FDK %
Sum3-81.60 * Kõ	0.48	0.14	5.54	0.79
Sumey-3	0.00	0.14	3.97	0.03
Sgv-NB * MM-Sum3	0.56	0.44	8.4	0.91
RSt * MM-NB,A	0.38	0.55	4.47	2.42
Wuhan2	1.02	2.21	2.46	4.85
RSt * MM-NB,B	2.11	2.94	23.77	12.52
Bence	6.85	3.45	21.77	26.85
Sgv-NB * MM-Sum3	0.72	3.94	5.12	2.45
Sum3-81.60 * Kõ	5.13	12.73	10.15	18.52
Kende	7.98	17.11	36.46	30.97
Sámson	8.99	17.36	46.9	35.00
Zu * Ré-NB	8.41	21.73	45.83	40.36
Jbj-50	11.61	23.97	33.34	45.82
Öthalom	10.94	24.47	41.35	58.27
Rába	12.83	25.68	31.6	47.76
Zugoly	12.13	26.39	35.37	50.06
Kalász	13.97	30.20	42.97	51.15
Góbé	13.12	40.55	46.09	52.55
Mean	6.51	14.11	24.76	26.74
LSD %	4.40	1.06	5.18	3.54

	FHB %	FDK inf. %	Yield red. %
FDK. %	0.9336***		
Yield red. %	0.8573***	0.9072***	
DON μg kg^{-1}	0.9435***	0.9799***	0.8838***

***$P = 0.001$.

Table 5. DON production following fungicides application in cultivar Zugoly, 1998. Data in mg kg^{-1}

Fungicide l/ha	Isolates				
	12377Fg	40Fg	12375Fc	12551Fc	Mean
Folicur Top+ Kolf.S. 1 + 1.5	10.04	4.47	5.74	13.29	8.39
Folicur Solo 1	7.61	1.16	10.34	16.64	8.94
Folicur Top 1	8.97	3.83	9.24	14.37	9.10
Falc. 1.0	12.88	5.84	10.57	17.69	11.75
Juwel 1	12.47	3.83	14.13	17.59	12.01
Falcon 0.8	12.32	6.74	17.76	19.62	14.11
Fusarium check	14.94	8.94	21.83	22.67	17.10
Amistar 1	20.10	12.31	25.13	26.29	20.96
Kolfugo S 1.5	14.72	5.53	26.67	37.56	21.12
Mean	11.40	5.46	13.97	19.51	12.59
LSD 5%					3.58

conidia do not contain DON. Mesterházy (1978) found no DON, but there were traces of diacetoxyscirpenol and neosolaniol and larger amount of diethylnivalenol in 14-day-old shaken cultures of *F. graminearum*, even though the isolates studied produced large quantities of DON after head inoculation. New data (Evans et al., 2000) confirms this, supporting the view that DON is synthesized only during pathogenesis.

Accordingly, DON has been considered to function as a pathotoxin. This led to the conclusion that selecting for toxin resistance directly or from calluses, somaclones or microspores on toxin-containing agar or culture filtrate-containing media would lead to the production of disease-resistant plants. Such work was started in China (Chen et al., 2000), Austria (Buerstmayr et al., 1996; Lemmens, 2000, pers. comm.), the Netherlands (Snijders 1997, pers. comm.) and Hungary. Positive results were achieved in China, where highly resistant plants were selected. However, later work revealed that among the surviving plantlets there were some disease-resistant plants, but there were also plants with considerable susceptibility. In Hungary (Ahmed et al., 1992; 1996) somaclones were screened on culture filtrate and T-2 toxin media. The surviving 10% of calluses were regenerated and followed to the R_3 generation allowing selection for *Fusarium* resistance. A parallel population was also added, where no toxin selection was applied and calluses were regenerated randomly. When the two populations were compared, it was found that in both populations there were plants with greater resistance than the parent cultivars, but there were also some plants with increased susceptibility. Monasterski (2000) found a close correlation between resistance and the regeneration ratio of calluses after selection on toxin-containing media, but the resistance of the regenerated progenies has not been tested. DON sensitivity of seedlings and resistance to disease correlated well (Liu et al., 1999) and seems, therefore, to confirm the role of resistance in suppressing DON activity, rather than to suggest that DON plays a major role in pathogenesis. Our conclusions are that neither DON nor T2 toxin are pathotoxins, and that toxin resistance and disease resistance are two different phenomena.

The effect of the cultivars on the DON-producing ability of given isolates should also be stressed. In susceptible and moderately susceptible cultivars, the disease-causing ability depends largely on DON-producing ability. However, in highly resistant genotypes this ability is strongly inhibited. Toxin production is also substrate-dependent. Under field conditions, a much narrower spectrum of toxins occurred in wheat than in rice grains in the laboratory (Szécsi and Bartók, 1996). The same phenomenon was recognized by Stack et al. (2000) and Miedaner et al. (2000). For this reason, toxin-producing ability should be measured under field conditions and to neutralize the cultivar effect, a set of genotypes with differing resistance should be used.

The level of DON concentration is also strongly influenced by the amount of precipitation following infection. Between years, ten-fold differences occurred when we considered the amount of FDK needed to produce 1 $\mu g\, g^{-1}$ DON. This means that aggressiveness and cultivar resistance are not alone in determining the final level of toxin concentration. In later phases of the disease, other mechanisms are important to control DON production. Fungicides play an important role in decreasing significantly the infection of cereals by Fusaria and their DON-producing capacity, although partially effective fungicides cause increased toxin contamination in parallel with a lower disease level.

The regulation of DON production during the development of FHB is a complex phenomenon. The aggressiveness of *F. graminearum* and *F. culmorum* correlates with production of DON and/or trichothecenes, suggesting that production of these toxins is an important component of aggressiveness. Other *Fusarium* species, which lack ability to produce DON, are poorly pathogenic. Resistant cultivars and fungicides can fully inhibit the disease-causing capacity of highly aggressive, DON-producing isolates. It seems that strategically, an increase in the availability of disease-resistant cultivars is the most important task required to combat the problems associated with toxin-producing *Fusarium* species.

References

Ahmed KZ, Mesterházy Á and Sági F (1996) *In vitro* production of *Fusarium* resistant wheat plants. In: Bajaj YPS (ed) Biotechnology in Agriculture and Forestry, Vol 36. Somaclonal Variation in Crop Improvement II (pp 3–19) Springer Verlag, Heidelberg

Ahmed KZ, Mesterházy Á and Sági F (1992) *In vitro* techniques for selecting wheat (*Triticum aestivum* L.) for *Fusarium*-resistance. I. Double-layer culture technique. Euphytica 57: 251–257

Alexander NJ, Proctor RH, McCormick SP and Plattner RD (1997) Generic and molecular aspects of the biosynthesis of trichothecenes by *Fusarium*. Cereal Research Communications 25: 315–320

Arseniuk E, Foremska E, Góral T and Chelkowski J (1999) *Fusarium* head blight reactions and accumulation of deoxynivalenol (DON) and some of its derivatives in kernels of wheat, triticale and rye. Journal of Phytopathology 147: 577–590

Bai G-H, Desjardins AE and Plattner RD (2000) Deoxynivalenol non-producing *Fusarium graminearum* causes initial infection but does not cause disease spread in wheat spikes. In: Proceedings of the International Symposium on Wheat Improvement for Scab Resistance (pp 224–233) Suzhou and Nanjing, China

Buerstmayr H, Lemmens M, Grausgruber H and Ruckenbauer P (1996) Breeding for scab resistance in wheat: Inheritance of resistance and possibilities for *in-vitro* selection. In: Dubin HJ, Gilchrist L, Reeves J and McNab A (eds) *Fusarium* Head Scab: Global Status and Future Prospects (pp 52–58) Mexico, D.F. CIMMYT

Chen L-F, Song YL and Xu YG (1996) Variation in the concentration of deoxynivalenol in the spikes of winter wheat infected by *Fusarium graminearum* Schw. Acta Phytopathologica Sinica 26: 25–28

Chen L-F, Bai G-H and Desjardins A (2000). Recent advances in wheat head scab research in China. USDA, Agricultural Research Service, On-line edition: www.scabusa.org. 63 pp

Desjardins A, Bai G-H, Plattner RD and Proctor RH (2000) Analysis of aberrant virulence of *Gibberella zeae* following transformation-mediated complementation of a trichithecene deficient (Tri5) mutant. Microbiology 146: 2059–2068

Desjardins A, Proctor RH, Bai G-H, McCormick SP, Shaner G, Buechley G and Hohn TM (1996) Reduced virulence of trichothecene antibiotic non-producing mutants of *Gibberella zeae* in wheat field tests. Molecular Plant Microbe Interaction 9: 775–781

Eudes F, Collin J, Rioux S and Comeau A (1997) The trichothecenes, a major component of wheat scab pathogenesis. Proceedings of the 5th European Fusarium Seminar. Cereal Research Communications 25: 95–496

Evans CK, Xie W, Dill-Macky R and Mirocha CJ (2000) Biosynthesis of deoxynivalenol in spikelets of barley inoculated with macroconidia of *Fusarium graminearum*. Plant Disease 84: 654–660

Gang GT, Miedaner T, Schuhmacher U, Schollenberger M and Geiger HH (1998) Deoxynivalenol and nivalenol production by *Fusarium culmorum* isolates of differing aggressiveness toward winter rye. Phytopathology 88: 879–884

Kang Z and Bucheauer H (1999) Immunocytochemical localization of Fusarium toxins in infected wheat spikes by *Fusarium culmorum*. Physiology of Molecular Plant Pathology 55: 275–288

Kang Z and Bucheauer H (2000) Ultrastructural and cytochemical studies on cellulose, xylan and pectin degradation in wheat spikes infected by *Fusarium culmorum*. Journal of Phytopathology 148: 263–275

Lamper Cs, Téren J, Bartók T, Komoróczy R, Mesterházy Á and Sági F (2000) Predicting DON contamination in *Fusarium*-infected wheat grains via determination of the ergosterol content. Cereal Research Communications 28: 337–344

Liu X-Q, Li X and Zhang X-M (1999) Relationship between resistance to scab and to *Fusarium graminearum* toxin in wheat varieties. Journal of Huazhong Agricultural University 18: 416–419

Mesterházy Á (1978) A *Fusarium graminearum* gomba szûrletének hatása búzára és kukoricára. (Effect of culture filtrates of *Fusarium graminearum* on wheat and corn). Növénytermelés 27: 11–20

Mesterházy Á (1983) Breeding wheat for resistance to *Fusarium graminearum* and *F. culmorum*. Z. Pflanzenzüchtung 91: 295–311

Mesterházy Á (1987) Selection of head blight resistant wheat through improved seedling resistance. Plant Breeding 98: 25–36

Mesterházy Á (1995) Types and components of resistance against *Fusarium* head blight of wheat. Plant Breeding 114: 377–386

Mesterházy Á (2001) Results of breeding for resistance against Fusarium head blight (FHB) in wheat. In: Proceedings of the 2001 National Fusarium Head Blight Forum (pp 254–258) Cincinnati

Mesterházy Á and Bartók T (1996) Control of *Fusarium* head blight of wheat by fungicide and its effect in the toxin contamination of the grains. Pflanzenschutz Nachrichten Bayer 49: 87–205

Mesterházy Á and Bartók T (1997) Effect of chemical control on FHB and toxin contamination of wheat. Cereal Research Communications 25: 1–783

Mesterházy Á and Bartók T (2001) Fungicide control of Fusarium head blight in wheat. In: Proceedings of the 2001 National Fusarium Head Blight Forum (pp 70–74) Cincinnati

Mesterházy Á, Bartók T, Mirocha CM and Komoróczy R (1999) Nature of resistance of wheat to *Fusarium* head blight and deoxynivalenol contamination and their consequences for breeding. Plant Breeding 118: 97–110

Miedaner T, Reinbrecht C and Schilling A (2000) Association among aggressiveness, fungal colonization, and mycotoxin production of 26 isolates of *Fusarium graminearum* in winter rye head blight. Zeitschrift f. Pflanzenkrankheiten u. Pflanzenschutz 107: 124–134

Miedaner T and Reinbrecht C (2001) Trichothecene content of rye and wheat genotypes inoculated with deoxynivalenol- and nivalenol-producing isolates of *Fusarium culmorum*. Journal Phytopathology 149: 245–251

Miller JD, Young JC and Sampson RD (1985) Deoxynivalenol and *Fusarium* head blight resistance in spring cereals. Phytopathologische Zeitschrift, 113: 359–367

Miller JD and Arnison PG (1986) Degradation of deoxynivalenol by suspension cultures of Fusarium head blight resistant wheat cultivar Frontana. Canadian Journal of Plant Pathology 8: 47–150

Mirocha CJ, Hui Yu, Evans CK, Kolaczkowski E and Dill-Macky R (1997) Chemistry and physiology of deoxynivalenol in pathogenesis. Cereal Research Communications 25: 309–313

Monastersky OA (2000) The use of an anther culture for selecting wheat varieties for resistance to ear fusariose. Russian Agricultural Science 2: 1–3

Moraru I, Raducanu F, Ittu M and Ciocazanu I (1998) *In vitro* reaction of some winter wheat (*Triticum aestivum* L.) and maize (*Zea mays* L.) genotypes to ZEN toxin. Annals of the Institute for Cereal Plant Technology, Fundulea 65: 29–35

Muthomo JW, Schütze A, Dehne HW, Mutitu EW and Oerke EC (2000) Characterisation of *Fusarium culmorum* isolates by mycotoxin production and aggressiveness to winter wheat. Zeitschrift Pflanzenkrankheiten u. Pflanzenschutz 107: 113–123

Parry DW, Jenkinson P and McLeod L (1995) *Fusarium* ear blight (scab) in small grain cereals – a review. Plant Pathology 44: 207–238

Proctor RH, Hohn TM and McCormick SP (1995) Reduced virulence of *Gibberella zeae* caused by disruption of a trichothecene toxin biosynthetic gene. Molecular Plant–Microbe Interactions 8: 593–601

Schroeder HW and Christensen JJ (1963) Factors affecting resistance of wheat to scab by *Gibberella zeae*. Phytopathology 53: 831–838

Snijders CHA (1994) Breeding for resistance to *Fusarium* in wheat and maize. In: Miller JD and Trenholm HL (eds) Mycotoxins in Grain Compounds other than Aflatoxin (pp 37–58) Eagan Press, St. Paul, MN

Stack RW, Frohberg RC and Casper H (1997) Reaction of spring wheats incorporating Sumai# 3 derived resistance to inoculation with seven *Fusarium* species. Proceedings of the 5th European *Fusarium* Seminar. Cereal Research Communications 25: 667–671

Stack RW, Wolf-Hall CE, Casper HH and Hansen JM (2000) DON level in grain from inoculated plants with *Fusarium graminearum* is not correlated to the DON producing potential of individual cultures 2000 National *Fusarium* Head Blight Forum, Cincinnati, 198 (Abstract)

Szécsi Á and Bartók T (1995) Trichothecene chemotypes of *Fusarium graminearum* isolated from corn in Hungary. Mycotoxin Research 11: 85–92

European Journal of Plant Pathology **108:** 685–690, 2002.

Relationship between growth and mycotoxin production by *Fusarium* species, biocides and environment

N. Magan, R. Hope, A. Colleate and E.S. Baxter
Applied Mycology Group, Cranfield Biotechnology Centre, Cranfield University, Silsoe, Bedford MK45 4DT, UK (Fax: +441525863540; E-mail: N.Magan@cranfield.ac.uk)

Key words: Fusarium species, water availability, fungicides, mycotoxins

Abstract

Fusarium head blight of cereals has, in recent years, become one of the most important pre-harvest diseases worldwide. This paper examines the *in vitro* efficacy of fungicides to control *Fusarium* species in cereals and the efficacy in the field on both *Fusarium* infection of ripening ears as well as their impact on mycotoxin production. Field studies suggest that fungicides such as tebuconazole and metconazole give good control of both *Fusarium* infection of ears and control of deoxynivalenol (DON) production. However, azoxystrobin and related fungicides are less effective, and grain from treated crops has sometimes been found to have increased concentrations of DON and nivalenol. Studies of isolates of *Fusarium culmorum* from different parts of Europe showed that complex interactions occur between environmental factors, fungicide type and isolate in relation to growth inhibition and DON production. These studies confirmed the ineffectiveness of azoxystrobin and suggest that environmental stress factors, particularly water availability and temperature, and low fungicide doses may stimulate mycotoxin production by Fusaria *in vitro* and in wheat grain.

Introduction

With the intensive cultivation of monoculture cereals, the input requirements, including pesticides and fungicides, needed to maximise yield have always been significant. In the last 10 years, the increased level of colonisation and infection by *Fusarium* spp., particularly of ripening ears of both temperate and tropical cereals has attracted much attention; firstly, because of the significant effects on yield and quality of harvested grain, and secondly, because of the ability of *Fusarium* spp. to produce a wide range of mycotoxins which can enter the human and animal food chains. Table 1 lists the key *Fusarium* spp. which can be involved in head blight and ear blight of cereals and their key mycotoxins which are the trichothecenes, zearelenone (ZEN), moniliformin and the fumonisins.

A significant effort has been concentrated on the development and use of fungicides for control of *Fusarium* spp. to prevent both infection and mycotoxin production. A significant amount of screening has been carried out on a range of systemic fungicides for control of *Fusarium* spp. in cereals, particularly wheat, barley and oats, where natural contamination with deoxynivalenol (DON) and nivalenol has been found.

Table 1. Summary of important *Fusarium* species and their mycotoxins

Species	Major mycotoxins
F. culmorum	Deoxynivalenol (DON), 3-acetyl DON, 15-acetyl DON, nivalenol, fusarenone (FX), Zearalenone (ZEN)
F. graminearum	DON, 15-ADON, NIV, FX, ZEN
F. sporotrichioides	T-2 toxin, HT-2 toxin, neosolaniol (NEO), diacetoxyscirpinol (DAS), FX, ZEN.
F. poae	T-2 toxin, HT-2 toxin, NIV, DAS, FX
F. verticilloides *F. proliferatum*	Fumonisins, moniliformin, fusarin C
F. oxysporum	Moniliformin, wortmannin, fusaric acid
F. sambucinum	Sambutoxin

For example, Snijders and Perowski (1990) showed a direct relationship between natural incidence of ear blight and DON contamination of wheat kernels by *F. culmorum*. Originally, a range of fungicides, such as mixtures based on the non-systemic prochloraz, and systemic fungicides such as tridemorph, carbendazim and zineb were used in the 1980s to control foliar and ear diseases (Moss, 1985; Moss and Frank, 1985; Magan and Lacey, 1986). The expanding acreage for cultivation of bread wheats in Europe in the 1980s also resulted in an increase in head blight of ripening ears by *Fusarium* spp., similar to that observed in Canada (Sutton, 1983; Parry et al., 1995). Complex interactions between species were also observed in Canada where early infection of ripening ears by *F. poae* and *F. avenaceum* were considered to facilitate later colonization by *F. culmorum* and *F. graminearum* (Sturz and Johnston, 1983). However, while information is available on the impact of environmental factors on the germination, growth and sporulation of *Fusarium* spp. from wheat (Magan and Lacey, 1984a), and *F. moniliforme* (=*F. verticillioides*) and *F. proliferatum* from maize (Marin et al., 1995), the relationship between these factors, fungicide efficacy and effects on growth and mycotoxin production have not been examined previously. This paper will consider information that is available, and present some recent studies where the impact of water availability, temperature and interactions with fungicides on DON production were examined for a range of *F. culmorum* strains from different parts of Europe (Italy, Norway, Sweden and the U.K.).

Relationship between fungicides, *Fusarium* spp. and mycotoxin production in cereals

Studies by Moss and Frank (1985) found that concentrations of tridemorph influenced production of T-2 toxin by *F. sporotrichioides*. At low concentrations (6–8 $\mu g\,ml^{-1}$), growth of *F. sporotrichoides* was enhanced while T-2 toxin and diacetoxyscirpenol (DAS) production was reduced. Higher fungicide concentrations (30–50 $\mu g\,ml^{-1}$) inhibited growth by >50% but stimulated formation of T-2 toxin five-fold. The same fungicide enhanced the production of aflatoxin by *Aspergillus flavus* (Badii and Moss, 1988). In contrast, the benzothiazole derivative, tricyclazole, completely inhibited aflatoxin production at concentrations only partially inhibiting growth (Fernando and Bean, 1986).

Treatment of *F. graminearum* infection of kernels of maize with maneb resulted in good control of both pathogen growth and ZEN production at 50 $\mu g\,ml^{-1}$ *in situ* and *in vitro* (Draughton and Churchville, 1983). More recently, Hasan (1993) compared the effects of dicloran, iprodione and vinclozolin and found that, at 500 $\mu g\,ml^{-1}$, the former eliminated DAS production and that 250 $\mu g\,ml^{-1}$ was sufficient to inhibit ZEN synthesis. In contrast, 500 $\mu g\,ml^{-1}$ of vinclozolin failed to prevent ZEN being produced by *F. graminearum*. Matthies and Buchenauer (1996) screened a range of fungicides in common use in Germany and found that of benomyl, thiabendazole, prochloraz, tebuconazole, tridemorph and fenopropimorph, the latter two had no effects on monoacetyl deoxynivalenol (3-AcDON) production in pure culture. The others inhibited production of the mycotoxins at 0.5–1.0 $\mu g\,ml^{-1}$. Tubiconazole induced a four-fold increase in the mycotoxin concentration at only 0.1 $\mu g\,ml^{-1}$. Similar results were obtained with *F. culmorum* at this concentration, whereas difenoconazole had no effect on growth, but increased the production of the same mycotoxin at 25°C, but not at 11°C (De Mello et al., 1998). D'Mello et al. (1998) also compared resistant and sensitive strains of *F. culmorum* to difenoconazole. Overall, the sensitive strains produced 3-AcDON in the presence of difenoconazole at 0.1 $\mu g\,ml^{-1}$, but less rapidly in resistant strains. It is possible that fungicides act as an additional stress factor stimulating synthesis of mycotoxins as a defence response. Recent studies have shown that metaconazole and tebuconazole not only inhibited growth but also caused marked morphological and cytochemical alterations of the hyphae when 20 $\mu g\,ml^{-1}$) fungicide was added to media (Kang et al., 2001a,b). They demonstrated that DON was localized in the cell walls, cytoplasm and mitochondria and vacuoles of the hyphae and that labelling density, using immunogold labelling, was significantly less dense than in untreated control hyphae of *F. culmorum*.

Very few experiments have considered interactions between fungicide concentration and environmental factors, particularly water availability and temperature, which to a large extent determine *Fusarium* ear blight of ripening ears. Surprisingly, no detailed studies of the efficacy of fungicides on mycotoxigenic *Fusarium* spp. have taken these abiotic interactions into account. This has been considered in studies with fumonisin-producing species and post-harvest preservatives such as propionic and sorbic acids (Marin et al., 1998; 1999).

Recent field studies were carried out to examine the efficacy of fungicide applications on *Fusarium* head blight of cereals and on mycotoxin production. For example, work by Milus and Parsons (1994) suggested that levels of mycotoxins may be increased by the application of fungicides. Ellner (2000) carried out field trials with azole fungicides (tebuconazole, metconazole) in two seasons in Germany and found that control of head blight and reductions in the levels of DON did not exceed 50%. Recent work by Jennings et al. (2000) and by Simpson et al. (2001) demonstrated complex interactions between the type of fungicide used and effects on colonization by *Fusarium* spp. and mycotoxin production. For example, tebuconazole was quite effective against *F. culmorum* and *F. avenaceum*, monitored both as number of diseased ears produced and using quantitative PCR to assess pathogen biomass. However, azoxystrobin had little effect on *F. culmorum* and *F. avenaceum*, but gave significant control of *Microdochium nivale* var. *majus*. Mixtures of azoxystrobin/prochloraz and azoxystrobin/fluquinconazole were less effective against *F. culmorum* than against *M. nivale*. Higher DON productions were found in plots treated with azoxystrobin (5.2 and 10.4 $\mu g\ ml^{-1}$ respectively) in both 1998 and 1999, and in 1999 with difenoconazole (11.2 $\mu g\ ml^{-1}$). They suggested that complex interactions occur between the differential impact of the fungicides, resulting in interactions between *M. nivale* and other *Fusarium* spp. which could result in this enhanced accumulation of DON. Siranidou and Buchenauer (2001) applied the fungicides two days prior to or two days post-inoculation with *F. culmorum* to ears of wheat and found that both tebuconazole and metconazole reduced disease on spikes of winter wheat by 60–70% and DON content by 50–70%. However, chlorothalonil, prochloraz and benomyl gave no disease control. Interestingly, azoxystrobin reduced disease incidence on the spikes but DON concentrations increased. Edwards et al. (2001) developed a quantitative PCR method to quantify trichothecene-producing *Fusarium* species based on primers from the trichodiene synthase gene (Tri 5). They were able to quantify a range of *Fusarium* head blight pathogens in relation to fungicide control. Metaconazole and tebuconazole gave good control of head blight and DON production, while azoxystrobin was ineffective. A good correlation was obtained between trichothcene-producing species and DON production in grain, but no correlation was found between *Fusarium* head blight and DON. They hypothesized that fungicides affected the *Fusarium* species in the head blight complex by altering the proportion of trichothcene-producing species in the infected ears and not by altering DON production. This method does, however, provide a tool for examining the interaction between *Fusarium* species and biotic and abiotic stress factors including fungicide applications.

Some studies have identified the environmental conditions which allow *F. culmorum* and *F. graminearum* to infect ripening ears of cereals (Jennings and Turner, 1996; Lacey et al., 1999; Savard et al., 2000). However, it is surprising that very few studies have examined the interaction between fungicides, environmental conditions and growth, and interactions between key *Fusarium* species. Furthermore, little is known of the variation in sensitivity between species of mycotoxigenic strains of *F. culmorum* in Europe. Few studies have examined the impact that interacting environmental conditions and fungicides might have on pathogen growth and mycotoxin production in infected plants and grain.

Impact of environmental factors on growth mycotoxigenic strains of *F. culmorum* from different parts of Europe

Recent studies in the Applied Mycology Group, Cranfield University, U.K. have examined the ecology of mycotoxigenic strains of *F. culmorum* from different parts of Europe. Studies have included a comparison of the water relations of the strains, assessments of the *in vitro* efficacy of fungicides used in cereals under different environmental regimes and their ability to control pathogen growth and mycotoxin production in wheat grain.

Initial studies showed that between strain differences (three strains each) from an individual country were, to a large extent, statistically non-significant (data not shown). Subsequent studies were carried out with one strain from each country. Figure 1 compares the effect of water activity (a_w) and temperature on representative strains of *F. culmorum* from Italy, Norway, Sweden and the U.K. Isolates behaved relatively similarly at 25 °C, with growth occurring down to 0.92 a_w. At 15 °C, growth was reduced by about 40–50%, but with a similar tolerance to a range of water availabilities. There were no significant differences between strains from the different countries at 25 °C but a strain from Norway appeared to grow significantly better than the others at 15 °C. These results, obtained on a milled wheat grain agar, were similar to those obtained earlier in studies comparing the water relations of *Fusarium* spp. from U.K. wheat (Magan and Lacey, 1984b).

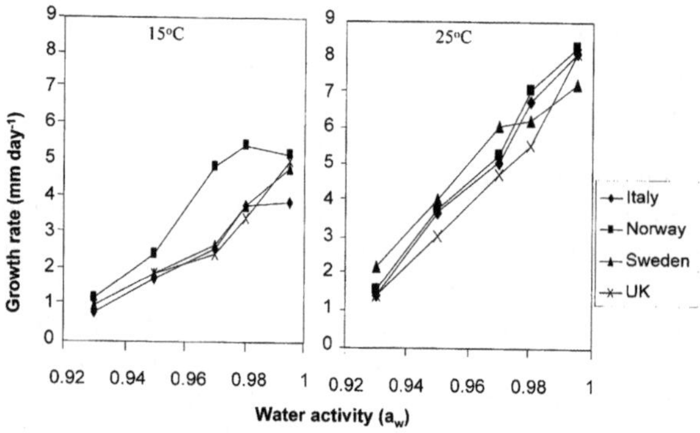

Figure 1. Comparison of water relations and growth of *F. culmorum* isolates from different European countries *in vitro* on a milled wheat agar medium.

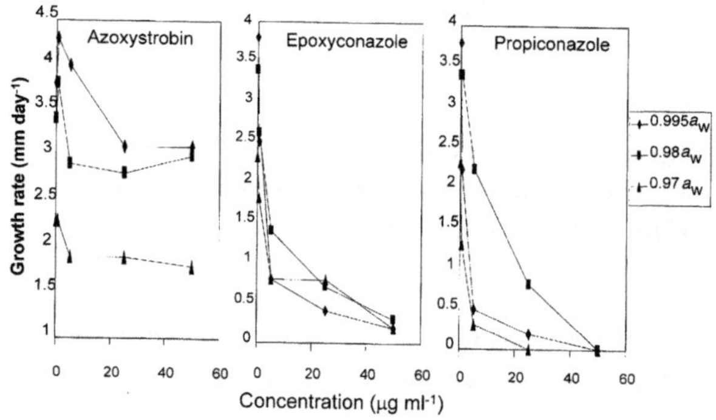

Figure 2. In vitro efficacy of three fungicides against a mycotoxigenic isolate of *F. culmorum* at 15 °C and three different water availability conditions (as indicated on the right axis).

The efficacy of different concentrations of three different fungicides used in cereals were compared at different a_w levels. Figure 2 shows that, for a U.K. isolate epoxyconazole and propiconazole were most effective at slightly reduced a_w levels with azoxystrobin being relatively ineffective, with little difference in efficacy at 20–50 μg ml^{-1}. Figure 3 further compares the effect of the fungicides at 15 °C on a strain of *F. culmorum* from each European country examined. This again shows that, in the range 0.5–50 μg ml^{-1} and 0.97 a_w, azoxystrobin was ineffective.

Subsequently, studies were carried out on irradiated wheat grain that had retained their germination capacity. Experiments were carried out on single layers of wheat grain modified with sterile water to the required water content (Lee and Magan, 2000), and with only the two most effective fungicides. A comparison of the growth rates across layers of wheat grain showed that

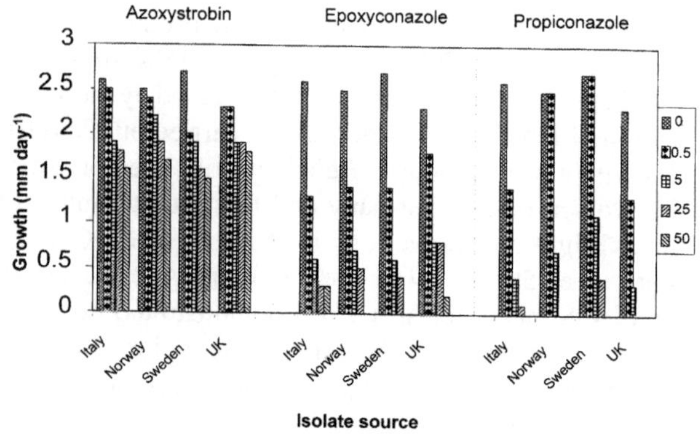

Figure 3. Comparison of the efficacy of three fungicides against growth of isolates of *F. culmorum* from four different European countries *in vitro* at 15 °C and 0.97 water activity. Concentrations of fungicides used are indicated on the right axis.

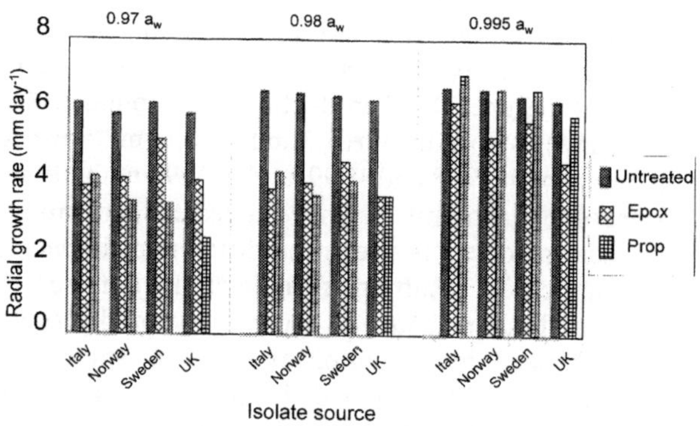

Figure 4. Effect of epoxyconazole (epox) and propiconazole (prop) at 25 μg ml^{-1} on growth of *F. culmorum* isolates at three different water availabilities on irradiated wheat grain at 25 °C.

at 25 μg ml^{-1} and freely available water, there was little effect of the fungicides on growth of the strains from the four different countries (Figure 4). However, at 0.98 and 0.97 a_w, propiconazole was more effective, reducing the growth rate by about 30–40% when compared to untreated controls under the same environmental conditions. Table 2 shows the effect of the fungicides on DON production by *F. culmorum*. The concentrations used show that there was a stimulation of DON production in the presence of the two fungicides. This points to a significant interaction between environmental conditions, especially water availability and temperature, fungicide type and concentration. To this must be added the interaction with other cereal ear mycoflora and the differential effect of fungicides on these fungal communities.

Table 2. Effect of water activity (a_w) and fungicide (25 μg ml^{-1}) on DON production by strains of *F. culmorum* from different parts of Europe grown on irradiated wheat grain with conserved germination capacity

Isolate source	a_w	Deoxynivalenol concentration (μg^{-1})		
		Control	Epoxiconazole	Propiconazole
U.K.	0.99	0.31	0.05*	0.55*
	0.98	0.72	0.88	1.23*
	0.97	1.09	1.31*	1.26*
Italy	0.99	0.67	0.41	0.88
	0.98	3.10	1.93	3.26
	0.97	5.03	20.00*	16.70*
Sweden	0.99	2.60	2.53	3.26
	0.98	19.33	5.93*	3.93
	0.97	5.03	20.00*	20.00*
Norway	0.99	3.20	3.40	2.26
	0.98	0.85	5.13*	4.73*
	0.97	1.16	20.00*	20.00*

*, indicates significantly different from the control at $P = 0.05$.

Conclusions

It is important that work now concentrates on some of the complex interactions between *Fusarium* infections, differential effects of fungicides, environmental factors and associated mycoflora which can occur on ripening ears of cereals. Furthermore, control of these fungi, and in particular, the prevention of increased concentrations of mycotoxins are essential if consumers are to be protected from toxic contaminants. Such effects are an important critical control point in the production process and more information is required for a hazard analysis critical control point system approach to preventing the entry of such mycotoxins into human and animal food chains.

Acknowledgements

In memory of Dr Wenche Langseth. We thank her, Dr Luciana Corazza, Dr Judith Turner, Dr Geoff Bateman and Dr. Hans Pettersson for supplying isolates and to the EU (QKL1-1999-00996) for supporting aspects of this work.

References

Badii F and Moss MO (1988) The effect of the fungicides tridemorph, fenpropimorph and fenarimol on growth and aflatoxin production by *Aspergillus parasiticus* Speare. Letters in Applied Microbiology 7: 37–39

D'Mello JPF, Macdonald AMC, Postel D, Dijiksman WTP, Dujardin A and Placinta C (1998) Pesticide use and mycotoxin production in *Fusarium* and *Aspergillus* phytopathogens. European Journal of Plant Pathology 104: 741–751

Draughton FA and Churchville DC (1985) Effect of pesticides on zearalenone production in culture and in corn plants. Phytopathology 75: 553–556

Edwards SG, Pirgozliev SR, Hare MC and Jenkinson P (2001) Quantification of trichothcene-producing *Fusarium* species in harvested grain by competitive PCR to determine efficacies of fungicides against *Fusarium* head blight of winter wheat. Applied and Environmental Microbiology 67: 1575–1580

Ellner FM (2000) Mycotoxins-quantification, control and regulation in agricultural produce. The 2000 BCPC Proceedings 75: 45–53

Fernando T and Bean G (1986) Effects of tricyclazole on growth, release of aflatoxin, and sterol and fatty acid content by *Aspergillus* isolates. Transactions of the British Mycological Society 87: 445–449

Hasan HAH (1993) Fungicide inhibition of aflatoxins, diacetoxyscerpinol and zearalenone production. Folia Micobiology 38: 295–298

Jennings P and Turner JA (1996) Towards the prediction of *Fusarium* ear blight epidemics in the UK-the role of humidity in disease development. The 1996 BCPC Crop Protection Conference – Pests and Diseases 1: 233–238

Jennings P, Turner JA and Nicholson P (2000) Overview of *Fusarium* ear blight in the UK – effect of fungicide treatment on disease control and mycotoxin production. The BCPC Crop Protection Conference – Pests and Diseases 6C-2: 707–712

Kang Z, Huang L and Buchenauer H (2001a) Ultrastructural and cytochemical studies of effects of the fungicide metconazole on *Fuarium culmorum in vitro* . Journal of Plant Diseases and Protection 108: 419–432

Kang Z, Huang L, Krieg U, Mauler-Machnik A and Buchenauer H (2001b) Effects of tebuconazole on morphology, structure, cell wall components and trichothecene production of *Fusarium culmorum in vitro*. Pest Management Science 57: 491–500

Lacey J, Bateman GL and Mirocha C (1999) Effects of infection time and moisture on development of ear blight and deoxynivalenol production by *Fusarium* spp. in wheat. Annals of Applied Biology 134: 277–283

Lee HB and Magan N (2000) Impact of environment and interspecific interactions between spoilage fungi and *Aspergillus ochraceus* and ochratoxin production on maize. International Journal of Food Microbiology 61: 11–16

Magan N and Lacey J (1984a) The effect of temperature and pH on the water relations of field and storage fungi. Transactions of the British Mycological Society 82: 71–81

Magan N and Lacey J (1984b) Water relations of some *Fusarium* species from infected wheat ears and grain. Transactions of the British Mycological Society 83: 281–285

Magan N and Lacey J (1986) The phylloplane microbial populations of wheat and effect of late fungicide applications. Annals of Applied Biology 109: 117–128

Marin S, Sanchis V and Magan N (1995) Water activity, temperature and pH effects on growth of *Fusarium moniliforme* and *F. proliferatum* isolates from maize. Canadian Journal of Microbiology 41: 1097–1105

Marin S, Magan N, Abellana M, Canela R, Ramos AJ and Sanchis V (1999) Selective effect of propionates on maize mycoflora and impact on fumonisin B1 accumulation. Journal of Stored Product Research 36: 203–214

Marin S, Sanchis V, Ramos AJ and Magan N (1998) Control of growth and fumonisin B1 production by *F. moniliforme* and *F. proliferatum* isolates in maize grain with propionate formulations. Food Additives and Contaminants 16: 555–563

Mathies A and Buchenauer H (1996) Investigations on the action of different active ingredients on the biosynthesis of mycotoxins in *Fusarium culmorum* and *F. graminearum*. In: Lyr H, Russell PE and Sisler HD (eds) Modern Fungicides and Antifungal Compounds (pp 199–204) Intercept Ltd., Andover

Milus EA and Parsons CE (1994) Evaluation of foliar fungicides for controlling *Fusarium* head blight of wheat. Plant Disease 78: 697–699

Moss MO (1985) Influence of agricultural biocides on mycotoxin formation in cereals. In: Chelkowski J (ed) Cereal Grain – Mycotoxins, fungi and quality in storage, Elsevier, Amsterdam

Moss MO and Frank M (1985) The influence of the fungicide tridemorph on T-2 toxin production by *Fusarium sporotrichioides*. Transactions of the British Mycological Society 84: 585–590

Parry DW, Jenkinson P and McLeod L (1995) *Fusarium* ear blight (scab) in small grain cereals: A review. Plant Pathology 44: 207–238

Savard ME, Sinha RC, Seaman WL and Fedak G (2000) Sequential distribution of the mycotoxin deoxynivalenol in wheat spikes after inoculation with *Fusarium graminearum*. Canadian Journal of Plant Pathology 22: 280–285

Simpson DR, Weston GE, Turner JA, Jennings P and Nicholson P (2001) Differential control of head blight pathogens of wheat by fungicides and consequences for mycotoxin contamination of grain. European Journal of Plant Pathology 107: 421–431

Siranidou E and Buchenauer H (2001) Chemical control of Fusarium head blight on wheat. Journal of Plant Diseases and Protection 108: 231–243

Snjiders CHA and Perowski J (1990) Effects of head blight caused by *Fusarium culmorum* on toxin content and weight of wheat kernels. Phytopathology 80: 566–570

Sturz AV and Johnston HW (1983) Early colonization of ears of wheat and barley by *Fusarium poae*. Canadian Journal of Plant Pathology 5: 107–110

Sutton JC (1982) Epidemiology of wheat head blight and maize caused by *F. graminearum*. Canadian Journal of Plant Pathology 4: 195–209

European Journal of Plant Pathology **108:** 691–698, 2002.

Genetic analysis of the role of trichothecene and fumonisin mycotoxins in the virulence of *Fusarium*

R.H. Proctor, A.E. Desjardins, S.P. McCormick, R.D. Plattner, N.J. Alexander and D.W. Brown
National Center for Agricultural Utilization Research, Agriculture Research Service, United States Department of Agriculture, 1815 North University Street, Peoria, IL 61604, USA (Fax: 3096816686; E-mail: proctorh@ncaur.usda.gov)

Key words: Fusarium graminearum, F. verticillioides, gene disruption

Abstract

The phytotoxicity of the *Fusarium* trichothecene and fumonisin mycotoxins has led to speculation that both toxins are involved in plant pathogenesis. This subject has been addressed by examining virulence of trichothecene and fumonisin-nonproducing mutants of *Fusarium* in field tests. Mutants were generated by transformation-mediated disruption of genes encoding enzymes that catalyze early steps in the biosynthesis of each toxin. Two economically important species of *Fusarium* were selected for these studies: the trichothecene-producing species *Fusarium graminearum*, which causes wheat head blight and maize ear rot, and the fumonisin-producing species *F. verticillioides*, which causes maize ear rot. Trichothecene-non-producing mutants of *F. graminearum* caused less disease than the wild-type strain from which they were derived on both wheat and maize, although differences in virulence on maize were not observed under hot and dry environmental conditions. Genetic analyses of the mutants demonstrated that the reduced virulence on wheat was caused by the loss of trichothecene production rather than by a non-target mutation induced by the gene disruption procedure. Although the analyses of virulence of fumonisin-non-producing mutants of *F. verticillioides* are not complete, to date, the mutants have been as virulent on maize ears as their wild-type progenitor strains. The finding that trichothecene production contributes to the virulence of *F. graminearum* suggests that it may be possible to generate plants that are resistant to this fungus by increasing their resistance to trichothecenes. As a result, several researchers are trying to identify trichothecene resistance genes and transfer them to crop species.

Introduction

The genus *Fusarium* produces numerous and chemically diverse mycotoxins. The harmful effects of many of these toxins on animal health are well documented, however, their effects on plant health are poorly understood. Because some *Fusarium* mycotoxins are phytotoxic and many *Fusarium* species are aggressive plant pathogens, it seems likely that some of these mycotoxins contribute to the ability of *Fusarium* to cause plant disease. Such a role for *Fusarium* mycotoxins in plant pathogenesis seems even more likely given the critical roles that fungal toxins play in the pathogenicity of fungi such as *Cochliobolus* (Yang et al., 1996).

At the National Center for Agricultural Utilization Research, work on mycotoxins and plant pathogenesis has focused on two groups of mycotoxins, the trichothecenes and the fumonisins. Although these toxins are produced by a number of *Fusarium* species, our efforts have dealt primarily with two economically important species: *F. graminearum*, which produces the trichothecenes deoxynivalenol (DON) and acetylated DON, and *F. verticillioides*, which produces fumonisins. Our general approach has been, first, to identify genes involved in early steps of mycotoxin biosynthesis, second, to generate mycotoxin-non-producing mutants of the fungi by specifically inactivating one of the biosynthetic genes, and third, to determine whether the mutants are less virulent than the mycotoxin-producing strains from which they were derived.

Trichothecenes and *Fusarium graminearum*

Fusarium graminearum (teleomorph *Gibberella zeae*) causes head blight of wheat and barley and ear rot of maize. Over the past decade severe head blight epidemics have occurred in North America, particularly in some Midwestern and eastern states of the US and in the Canadian provinces of Manitoba, Ontario, and Québec. It has been estimated that during the 1990s head blight caused losses in excess of $3.5 billion (US) in the USA and Canada (Windels, 2000). These losses result from reduced quality and yield of grain and from contamination of grain with trichothecenes, particularly DON. Trichothecenes are of concern because they inhibit protein synthesis and as a result can induce an array of health problems in animals and humans (Joffe, 1986).

Biochemistry and genetics of biosynthesis

Trichothecenes are sesquiterpenoids and their biosynthesis diverges from general isoprenoid metabolism with the formation of trichodiene via the cyclization of farnesyl pyrophosphate (Figure 1). Trichodiene undergoes a series of oxygenations, cyclizations, isomerizations, and esterifications to yield bioactive trichothecenes such as DON and acetylated DON (Desjardins et al., 1993). Over the past decade, the genes encoding enzymes that catalyze most of these steps have been identified and found to be located within a gene cluster (Hohn et al., 1993; Keller and Hohn, 1997). At least eight genes form the trichothecene biosynthetic gene cluster in *F. graminearum* and in the closely related species *F. sporotrichioides* (Brown et al., 2001). These genes include *TRI5*, which encodes trichodiene synthase (Hohn and Beremand, 1989); *TRI3*, which encodes an acetylase (McCormick et al., 1996); *TRI4* and *TRI11*, which encode cytochrome P450 mono-oxygenases (Alexander et al., 1998; Hohn et al., 1995); *TRI6* which encodes a transcription factor (Hohn et al., 1999), and *TRI12*, which encodes a major facilitator transport protein (Alexander et al., 1999a). At least, one trichothecene biosynthetic gene, *TRI101*, is located outside the trichothecene gene cluster. The *TRI101* protein acetylates the oxygen attached to carbon atom 3 (C-3) of trichothecenes and this markedly reduces their toxicity (Alexander et al., 1999b; Kimura et al., 1998; McCormick et al., 1999).

Trichothecene production and virulence

To examine the role of trichothecenes in the virulence of *F. graminearum* in wheat head blight and maize ear rot, the *TRI5* gene was inactivated via transformation-mediated gene disruption. We chose to disrupt *TRI5* because it encodes the enzyme that catalyzes the first committed reaction in trichothecene biosynthesis, that is, the cyclization of farnesyl pyrophosphate to trichodiene (Figure 1). Thus, disruption of *TRI5* results in mutants that cannot carry out this first biosynthetic step and therefore cannot produce DON, acetylated DON or any trichothecene biosynthetic intermediates (Proctor et al., 1995). To disrupt *TRI5*, we employed a standard protoplast transformation protocol with the hygromycin resistance gene, *HygB*, as a selectable marker (Proctor et al., 1995). Several types of disruption mutants were obtained. However, only two mutants were used in most analyses (Desjardins et al., 1996). The first disruption mutant, GZT33, had two incomplete copies of the *TRI5* coding region, one of which was interrupted by the *HygB* marker. As a result, the *TRI5* gene was inactive in this mutant. The second mutant, GZT40, had a single copy of the *TRI5* coding region that was interrupted by *HygB* and as a result was also inactive.

In wheat field tests, the *TRI5*-disrupted, trichothecene-non-producing mutants caused less severe head blight symptoms and smaller reductions in yield than the trichothecene-producing progenitor strain from which they were derived (Figure 2) (Desjardins et al., 1996). In addition, grain harvested from plants inoculated with the mutants had lower levels of infection and higher rates of germination than seed from plants inoculated with the producing progenitor strain. The reduced virulence of the mutants was observed over several years, at different locations (Illinois and Indiana), and in six different wheat cultivars (Desjardins et al., 1996). These data provide strong evidence that trichothecene production contributes to the ability of *F. graminearum* to cause wheat head blight.

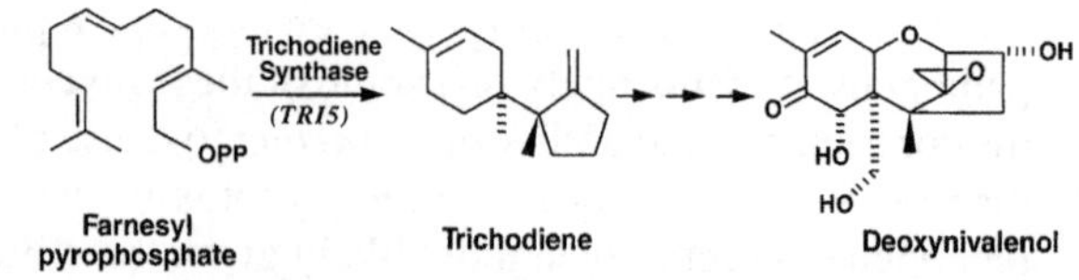

Figure 1. The biosynthesis of the trichothecene deoxynivalenol (DON) from the isoprenoid intermediate farnesyl pyrophosphate via the trichothecene parent compound, trichodiene.

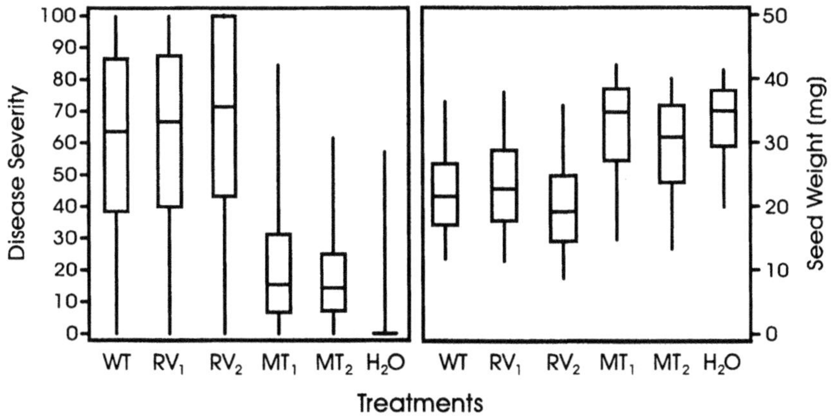

Figure 2. Head blight severity and yield (seed weight) of wheat inoculated with trichothecene producing and non-producing strains of *F. graminearum*. The data are presented as Box-Whisker plots. A box represents the range in which 50% of the data lie, the horizontal line within each box represents the median, and the vertical lines represent the range within which 95% of the data lie. WT, trichothecene-producing progenitor strain; RV_1 and RV_2, trichothecene-producing *TRI5* revertant strains; MT_1 and MT_2, trichothecene-non-producing *TRI5* disruption mutants GZT33 and GZT40; and H_2O, water inoculation. Disease severity was measured as the percent blighted spikelets per head at 19 days after inoculation, and seed weight was measured as the weight of seeds from individual heads. This figure was adapted from (Desjardins et al., 1996).

In a field experiment conducted with two maize cultivars in Ontario, Canada, the trichothecene-non-producing mutants caused significantly less severe ear rot symptoms and lower reductions in grain yield than the trichothecene-producing progenitor strain (Figure 3) (Harris et al., 1999). In the same experiment conducted in Illinois, the mutants induced less severe symptoms than the wild-type strains on one cultivar, but not on the other. In addition, reductions in grain yield caused by mutant and wild-type strains did not differ significantly in Illinois (Harris et al., 1999). The lack of differences in disease induced by the mutant and wild-type strains in Illinois was attributed to relatively low levels of disease induced by the wild-type strain under the hot and dry environmental conditions that prevailed during the experiment. In Ontario, environmental conditions were cooler and the wild-type strain consistently caused high levels of disease. Despite the experimental problems in Illinois, the maize field experiments indicate that trichothecene production contributes to the ability of *F. graminearum* to cause maize ear rot.

Non-target effects of transformation on virulence

The protoplast transformation protocol that was used to disrupt *TRI5* can induce mutations in genes other than those targeted for disruption, and it is possible that the reduced virulence of *TRI5* disruption mutants was caused by such a non-target mutation(s) rather than the inability of the mutants to produce trichothecenes (Desjardins et al., 2000a; Proctor et al., 1997). To address this concern, three types of experiments were done: a reversion analysis, a genetic analysis, and a complementation analysis. In the reversion analysis, the self-fertility of *F. graminearum* was exploited to induce the disrupted *TRI5*, in mutant GZT33, to revert to wild type (Proctor et al., 1997). This reversion was possible because the disrupted *TRI5* in GZT33 included two incomplete copies of the *TRI5* coding region. During meiosis, homologous regions of these two copies could recombine to generate a wild-type *TRI5* coding region and this, in turn, would restore trichothecene production. One progeny obtained by self-fertilizing GZT33 was found to have the wild-type *TRI5* gene and to be restored to trichothecene production. When tested in wheat head blight and maize ear rot assays, this revertant progeny was significantly more virulent than GZT33 and was as virulent as the original wild-type progenitor strain from which GZT33 was derived (Figures 2 and 3) (Desjardins et al., 1996; Harris et al., 1999). The fact that the restoration of trichothecene production restored high virulence in *F. graminearum* indicates that the low virulence of the *TRI5* disruption mutant GZT33 was due to its inability to produce trichothecenes rather than a non-target effect of the transformation procedure.

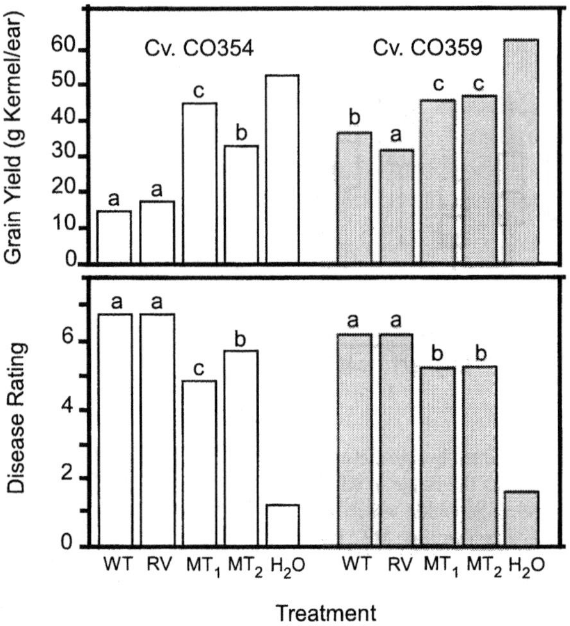

Figure 3. Mean ear rot ratings and grain yield for the two maize cultivars CO354 and CO359 grown in Ontario and inoculated with trichothecene producing and non-producing strains of *F. graminearum*. WT, trichothecene-producing progenitor strain; RV, trichothecene-producing *TRI5* revertant strain; MT_1 and MT_2, trichothecene-non-producing *TRI5* disruption mutants GZT33 and GZT40; and H_2O, water inoculation. A semi-subjective scale was used to measure ear rot ratings (Harris et al., 1999). For a given combination of cultivar and disease measurement (yield or symptoms), bars labeled with different letters are significantly different ($P < 0.05$). Data were taken from Harris et al. (1999).

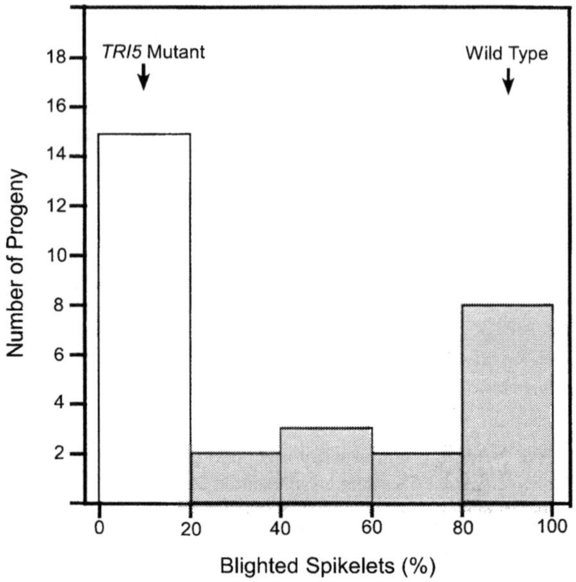

Figure 4. Distribution of head blight symptoms among trichothecene-producing (shaded) and trichothecene-non-producing (not shaded) progeny from a cross between *TRI5* disruption mutant GZT40 and a wild-type strain. Arrows indicate the level of head blight caused by the parental strains. Data were taken from Desjardins et al. (2000).

In the genetic analysis to assess the potential relationship of non-target mutation(s) and reduced virulence, disruption mutant GZT40 was crossed with a trichothecene-producing, highly virulent strain of *F. graminearum* (Desjardins et al., 2000a). Strain GZT40 was chosen for the genetic analysis because the structure of its disrupted *TRI5* allele was such that it could not revert to the wild-type allele. Also, GZT40 and the wild-type parental strain were genetically marked with nitrate-non-utilizing mutations to facilitate the selection of progeny that resulted from a cross between the two parents rather than from self-fertilization. The progeny of this cross were analyzed to determine whether they had the wild-type or disrupted *TRI5* allele, whether they produced trichothecenes, and whether they had high or low virulence on wheat heads. All progeny that had the disrupted *TRI5* allele did not produce DON and exhibited low virulence on wheat heads, whereas progeny with the wild-type *TRI5* allele produced DON and exhibited high virulence (Figure 4) (Desjardins et al., 2000a).

In the complementation analysis to assess the possible relationship of non-target mutation(s) and reduced virulence, a functional *TRI5* allele (*TRI5-4*) was transformed into *TRI5* disruption mutant GZT40 (Proctor et al., 1997). Although the *TRI5-4* allele restored trichothecene production to the mutant it did not restore high virulence on wheat heads (Desjardins et al., 2000a). However, a subsequent genetic analysis revealed that the low virulence of the *TRI5-4*-complemented GZT40 was caused by a non-target mutation that had been induced when the *TRI5-4* allele was transformed into GZT40 rather than by the initial transformation used to disrupt *TRI5* (Desjardins et al., 2000a).

The results from the reversion, genetic, and complementation and subsequent genetic analyses were all consistent and indicated that the reduced virulence of the *TRI5* disruption mutants was not caused by a non-target mutation from the transformation process used to disrupt *TRI5*. Thus, the three analyses confirmed that the reduced virulence of the *TRI5* disruption mutants was caused by their inability to produce trichothecenes.

Fumonisins and *Fusarium verticillioides*

Fusarium verticillioides (syn. *F. moniliforme*, teleomorph *G. moniliformis*, syn *G. fujikuroi* mating population A) causes ear and stalk rot of maize, but it is also frequently present in healthy maize tissue (Munkvold and Desjardins, 1997). Although, the fungus does not typically cause severe epidemics, it is of considerable importance because of its almost ubiquitous distribution in maize-growing regions of North America combined with its ability to produce fumonisins. Fumonisins cause the formation of cancerous tumors in laboratory rodents and there is an epidemiological correlation between the consumption of fumonisin-contaminated grain and human esophageal cancer in some regions of the world (Howard et al., 1999; Nelson et al., 1993). Fumonisins also cause several fatal livestock diseases, such as lung edema in swine and leukoencephalomalacia in horses (Nelson et al., 1993). At the cellular level, fumonisins disrupt sphingolipid metabolism and, because sphingolipids function in a number of cellular processes, it is likely that this disruption leads to the various fumonisin-induced animal diseases (Howard et al., 1999; Wang et al., 1991). Given their carcinogenicity and their widespread occurrence in maize, fumonisins have the potential to severely impact maize producers in North America.

Biochemistry and genetics of biosynthesis

Most strains of *F. verticillioides* produce fumonisin B_1 (FB_1) in highest abundance with smaller amounts of the less oxygenated fumonisins, fumonisins B_2 (FB_2) and B_3 (FB_3) and B_4 (FB_4) (Figure 5) (Nelson et al., 1993). Precursor-feeding experiments with labeled acetate indicate that these fumonisins are synthesized from a linear 20-carbon polyketide that undergoes carbonyl reduction, hydroxylation, condensation with alanine, and esterification with two tricarboxylic acids (Blackwell et al., 1994; Proctor et al., 1999a). The order in which the hydroxylations, condensation, and esterification occur is not clear. However, feeding studies with FB_2, FB_3, and FB_4 indicate that hydroxylation of C-10 must occur before hydroxylation of C-5 in the formation of FB_1 (Proctor et al., 1999a). In addition, the presence of a methyltransferase-encoding region in a polyketide synthase gene (see below) required for fumonisin biosynthesis indicates that the methyl groups at C-12 and C-16 of fumonisins are added during the synthesis of the polyketide (Figure 5).

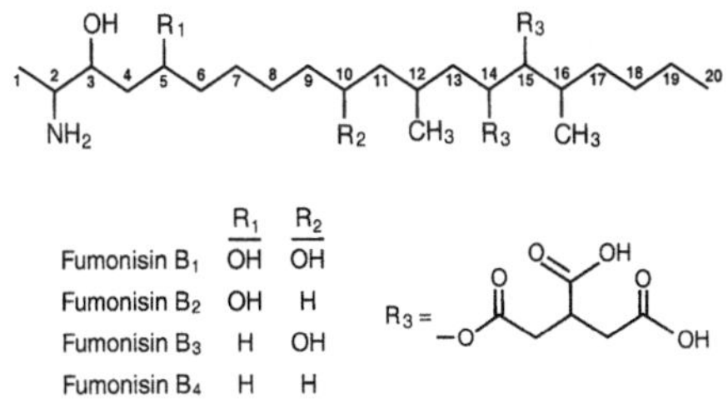

Figure 5. Structures of Fumonisins B_1, B_2, B_3, and B_4.

The fumonisin polyketide synthase gene was identified via PCR using degenerate polyketide synthase primers and a cDNA template that was prepared from a fumonisin-producing liquid culture of *F. verticillioides* (Proctor et al., 1999b). The amplification product from this PCR was used to isolate an 8-kb gene (*FUM5*) that encodes a polyketide synthase. Disruption of *FUM5* in *F. verticillioides*, via essentially the same method used to disrupt *TRI5* in *F. graminearum*, reduced fumonisin production by over 99% and thus demonstrated that *FUM5* participates in fumonisin biosynthesis (Proctor et al., 1999b). It is hypothesized that the polyketide synthase encoded by *FUM5* synthesizes the putative 20-carbon linear polyketide that forms C-3 through C-20 of the fumonisin backbone.

FUM5 is the first fumonisin biosynthetic gene to be characterized at the molecular level and presents an opportunity to begin critical studies on the molecular genetics and biochemistry of fumonisin biosynthesis as well as studies on the role of these mycotoxins in the ecology of *F. verticillioides*. Already, sequence analysis of regions up and down stream of *FUM5* have revealed the presence of other genes. Disruption and expression analyses have revealed that four genes (*FUM6*, *FUM7*, *FUM8*, and *FUM9*) immediately downstream of *FUM5* also participate in fumonisin biosynthesis (Seo et al., 2001). These data, along with those from classical genetic analyses of natural variants of *F. verticillioides* with altered fumonisin production phenotypes (Desjardins et al., 1996), indicate that fumonisin biosynthetic genes are clustered in a manner similar to trichothecene biosynthetic genes.

Fumonisin production and virulence

FUM5 is analogous to *TRI5* in that it encodes an enzyme that catalyzes an early step in mycotoxin biosynthesis. As a result, *FUM5* disruption mutants should be useful tools in the analysis of the role of fumonisins in the virulence of *F. verticillioides* on maize. Although these analyses are not complete, preliminary results indicate that *FUM5* disruption mutants cause high levels of ear rot, similar to those caused by the fumonisin-producing strains from which the mutants were derived (Desjardins et al., 2000b). These preliminary results are consistent with another set of experiments in which a natural variant of *F. verticillioides* that does not produce fumonisins caused the same levels of maize ear rot as a wild-type fumonisin producing strain (Desjardins and Plattner, 2000). Thus, all results to date indicate that fumonisin production does not have a major effect on the ability of *F. verticillioides* to cause maize ear rot.

Mycotoxin resistance genes and plant disease control

One of the reasons to study the role of mycotoxins in plant pathogenesis is to identify weak links in the producing fungi that can be exploited to control plant disease and mycotoxin contamination problems. The finding that trichothecenes contribute to the virulence of *F. graminearum* has revealed such a weak link. That is, it may be possible to incorporate trichothecene resistance into wheat and maize to render these crops resistant to diseases caused by *F. graminearum*. Reduction of these diseases should, in turn, result in less trichothecene contamination in the crops. Already, researchers have identified three genes that confer resistance to trichothecene: the *Fusarium TRI101* gene, which encodes an enzyme that reduces the toxicity of trichothecenes (Kimura et al., 1998; McCormick et al., 1999); the rice *Rpl3* gene, which encodes the ribosomal protein L3 and the wild-type form of which is a likely target of trichothecenes (Harris and Gleddie, 2000; 2001); and the yeast *PDR5* gene, which encodes an ABC transporter (Balzi et al., 1994). Researchers have already shown that these genes confer trichothecene resistance when expressed in tobacco (Harris and Gleddie, 2001; Muhitch et al., 2000). Whether such genes can confer resistance to trichothecenes and *F. graminearum* in wheat and maize is currently under investigation (Okubara et al., 2000).

If fumonisin production does not have a major effect on the ability of *F. verticillioides* to cause maize ear rot, as preliminary results suggest, it is unlikely that fumonisin resistance would be an effective way to control this disease or fumonisin contamination in maize. However, *F. verticillioides* produces several other toxins (e.g. fusaric acid and the fusarins) and probably other factors (Desjardins and Proctor, 1999) that may contribute to the virulence of this fungus. It is possible that methods to control *F. verticillioides*-induced maize ear rot and fumonisin contamination will become apparent if the relationships of these toxins and virulence are critically examined.

Conclusions

Analyses of *TRI5* disruption mutants have revealed that trichothecene production contributes to high levels of virulence of *F. graminearum* on wheat and maize. How the toxins enhance virulence of *F. graminearum* is not clear, however, it seems likely that the inhibitory effects of trichothecenes on protein synthesis could impair plant defenses. The analyses also indicated that trichothecenes are virulence factors rather than pathogenicity factors. That is, trichothecenes enhance the severity of disease caused by *F. graminearum*, but in the absence of their production the fungus can still cause low levels of disease. The low levels of disease caused by the trichothecene-non-producing mutants demonstrate that other factors contribute to the ability of the *F. graminearum* to cause disease.

If fumonisins are not involved in maize ear rot, it is possible that they enhance the virulence of *F. verticillioides* on some other host plant(s). Such a situation exists with trichothecenes, which enhance the virulence of *F. sporotrichioides* and *F. sambucinum* on parsnip root but not on potato tubers (Desjardins et al., 1993). Furthermore, the lack of a role for fumonisins in virulence does not preclude a role for them in some other aspect of the ecology of *F. verticillioides*.

While transformation-mediated gene disruption has been a crucial tool in studying the role of mycotoxins in plant diseases, one drawback of the technique is that the transformation process can induce non-target mutations that reduce virulence. Thus, gene disruption should not always be considered sufficient to demonstrate the function of a gene involved in complex physiological processes such as virulence. As described above, the presence of non-target mutations can be assessed by reversion, genetic and complementation analyses, although caution must be exercised with transformation-mediated complementation analyses because they can also induce non-target mutations. Alternative transformation protocols, such as gene gun and *Agrobacterium*-mediated methods, are currently being developed to disrupt genes in fungi. These methods may reduce or eliminate non-target mutations because they employ intact cells rather than protoplasts and, therefore, may be less stressful to fungi.

If trichothecene production contributes to the virulence of *F. graminearum*, it follows that plants that are resistant to the toxins should also be resistant to the fungus. Thus, studying the role of trichothecenes in virulence has revealed a strategy that could control the crop diseases and the resulting mycotoxin contamination problems caused by *F. graminearum*. This should lend support for future experiments to examine the role of other mycotoxins and other potential virulence factors in plant diseases caused by *Fusarium*.

References

Alexander NJ, Hohn TM and McCormick SP (1998) The *TRI11* gene of *Fusarium sporotrichioides* encodes a cytochrome P450 monooxygenase required for C-15 hydroxylation in trichothecene biosynthesis. Applied and Environmental Microbiology 64: 221–225

Alexander NJ, Hohn TM and McCormick SP (1999a) TRI12, a trichothecene efflux pump from *Fusarium sporotrichioides*: Gene isolation and expression in yeast. Molecular and General Genetics 261: 977–984

Alexander NJ, McCormick SP and Ziegenhorn SL (1999b) Phytotoxicity of selected trichothecenes using *Chlamydomonas reinhardtii* as a model system. Natural Toxins 7: 265–269

Balzi E, Wang M, Leterme S, van Dyck L and Goffeau A (1994) PDR5, a novel yeast multidrug resistance conferring transporter controlled by the transcription regulator PDR1. Journal of Biological Chemistry 269: 2206–2214

Blackwell BA, Miller JD and Savard ME (1994) Production of carbon 14-labeled fumonisin in liquid culture. Journal of AOAC International 77: 506–511

Brown DW, McCormick SP, Alexander NJ, Proctor RH and Desjardins AE (2001) A Genetic and biochemical approach to study trichothecene diversity in *Fusarium sporotrichioides* and *Fusarium graminearum*. Fungal Genetics and Biology 32: 121–133

Desjardins AE, Bai G, Plattner RD and Proctor RH (2000a) Analysis of aberrant virulence of *Gibberella zeae* following transformation-mediated complementation of a trichothecene-deficient (*Tri5*) mutant. Microbiology 146: 2059–2068

Desjardins AE, Hohn TM and McCormick SP (1993) Trichothecene biosynthesis in *Fusarium* species: Chemistry, genetics, and significance. Microbiological Reviews 57: 595–604

Desjardins AE and Plattner RD (2000) Fumonisin B1-nonproducing strains of *Fusarium verticillioides* cause maize (*Zea mays*) ear infection and ear rot. Journal of Agricultural and Food Chemistry 48: 5773–5780

Desjardins AE, Plattner RD and Proctor RH (1996) Linkage among genes responsible for fumonisin biosynthesis in *Gibberella fujikuroi* mating population A. Applied and Environmental Microbiology 62: 2571–2576

Desjardins AE, Plattner RD and Proctor RH (2000b) Maize ear rot and systemic infection by fumonisin B1-nonproducing mutants of *Gibberella fujikuroi* mating population A. Phytopathology 90 (Suppl): 596

Desjardins AE and Proctor RH (1999) Biochemistry and genetics of *Fusarium* toxins. In: B Summerell B, Leslie JF, Backhouse D, Bryden WL and Burgess LW (ed) Fusarium: Paul E. Nelson Symposium APS Press, St. Paul, USA.

Desjardins AE, Proctor RH, Bai G, McCormick SP, Shaner G, Buechley G and Hohn TM (1996) Reduced virulence of trichothecene antibiotic-non-producing mutants of *Gibberella zeae* in wheat field tests. Molecular Plant–Microbe Interactions 9: 775–781

Harris LJ, Desjardins AE, Plattner RD, Nicholson P, Butler G, Young JC, Weston G, Proctor RH and Hohn TM (1999) Possible role of trichothecene mycotoxins in virulence of *Fusarium graminearum* on maize. Plant Disease 83: 954–960

Harris LJ and Gleddie SC (2000) Developing plant tolerance to trichothecene mycotoxins using a transgenic approach. Phytopathology 90 (Suppl): 596

Harris LJ and Gleddie SC (2001) A modified *Rpl3* gene from rice confers tolerance of the *Fusarium graminearum* mycotoxin deoxynivalenol to transgenic tobacco. Physiological and Molecular Plant Pathology 58: 173–181

Hohn TM and Beremand P (1989) Isolation and nucleotide sequence of a sesquiterpene cyclase gene from the trichothecene-producing fungus *Fusarium sporotrichioides*. Gene 79: 131–138

Hohn TM, Desjardins AE and McCormick SP (1995) The *Tri4* gene of *Fusarium sporotrichioides* encodes a cytochrome P450 involved in trichothecene biosynthesis. Molecular and General Genetics 248: 95–102

Hohn TM, Krishna R and Proctor RH (1999) Characterization of a transcriptional activator controlling trichothecene toxin biosynthesis. Fungal Genetics and Biology 26: 224–235

Hohn TM, McCormick SP and Desjardins AE (1993) Evidence for a gene cluster involving trichothecene-pathway biosynthetic genes in *Fusarium sporotrichioides*. Current Genetics 24: 291–295

Howard PC, Eppley RM, Stack ME, Warbritton A, Voss KA, Lorentzen RJ, Kovach R and Bucci TJ (1999) Carcinogenicity of fumonisin B_1 in a two-year bioassay with Fischer 344 rats and B6C3F$_1$ mice. Mycotoxins Supplement 1999: 45–54

Joffe A (1986) *Fusarium* Species: Their Biology and Toxicology. John Wiley and Sons, New York

Keller NP and Hohn TM (1997) Metabolic pathway gene clusters in filamentous fungi. Fungal Genetics and Biology 21: 17–29

Kimura M, Kaneko I, Komiyama M, Takatsuki A, Koshino H, Yoneyama K and Yamagushi I (1998) Trichothecene 3-*O*-acetyltransfease protects both the producing organism and transformed yeast from related mycotoxins. Journal of Biological Chemistry 273: 1654–1661

McCormick SP, Alexander NJ, Trapp SC and Hohn TM (1999) Disruption of *TRI101*, the gene encoding trichothecene 3-*O*-acetyltransferase, from *Fusarium sporotrichioides*. Applied and Environmental Microbiology 65: 5252–5256

McCormick SP, Hohn TM and Desjardins AE (1996) Isolation and characterization of *Tri3*, a gene encoding 15-*O*-acetyltransferase from *Fusarium sporotrichioides*. Applied and Environmental Microbiology 62: 353–359

Muhitch MJ, McCormick SP, Alexander NJ and Hohn TM (2000) Transgenic expression of the *TRI101* or *PDR5* gene increases resistance of tobacco to the phytotoxic effects of the trichothecene 4,15-diacetoxyscirpenol. Plant Science 157: 201–207

Munkvold GP and Desjardins AE (1997) Fumonisins in maize: Can we reduce their occurrence? Plant Disease 81: 556–565

Nelson PE, Desjardins AE and Plattner RD (1993) Fumonisins, mycotoxins produced by *Fusarium* species: Biology, chemistry, and significance. Annual Review of Phytopathology 31: 233–252

Okubara PA, Hohn TM, Berka RM, Alexander NA, Wang Z, Hart LP and Blechl AE (2000) Optimizing the expression of candidate anti-*Fusarium* protein genes in hexaploid wheat. Proceedings of the 2000 National Fusarium Head Blight Forum pp 39–43

Proctor RH, Desjardins AE and Plattner RD (1999a) Biosynthetic and genetic relationships of B-series fumonisins produced by *Gibberella fujikuroi* mating population A. Natural Toxins 7: 251–258

Proctor RH, Desjardins AE, Plattner RD and Hohn TM (1999b) A polyketide synthase gene required for biosynthesis of fumonisin mycotoxins in *Gibberella fujikuroi* mating population A. Fungal Genetics and Biology 27: 100–112

Proctor RH, Hohn TM and McCormick SP (1995) Reduced virulence of *Gibberella zeae* caused by disruption of a trichothecene toxin biosynthetic gene. Molecular Plant–Microbe Interactions 8: 593–601

Proctor RH, Hohn TM and McCormick SP (1997) Restoration of wild-type virulence to *Tri5* disruption mutants of *Gibberella zeae* via gene reversion and mutant complementation. Microbiology 143: 2583–2591

Seo J-A, Proctor RH and Plattner RD (2001) Characterization of four clustered and coregulated genes associated with fumonisin biosynthesis in *Fusarium verticillioides*. Fungal Genetics and Biology 34: 155–165

Wang E, Norred WP, Bacon CW, Riley RT and Merrill AH (1991) Inhibition of sphingolipid biosynthesis by fumonisins: Implications for diseases associated with *Fusarium moniliforme*. Journal of Biological Chemistry 266: 14486–14490

Windels C (2000) Economic and social impacts of Fusarium Head Blight: Changing farms and rural communities in the northern Great Plains. Phytopathology 90: 17–21

Yang G, Rose MS, Turgeon BG and Yoder OC (1996) A polyketide synthase is required for fungal virulence and production of the polyketide T-toxin. The Plant Cell 8: 2139–2150

European Journal of Plant Pathology **108:** 699–703, 2002.

Saccharomyces cerevisae and *Arabidopsis thaliana*: Useful model systems for the identification of molecular mechanisms involved in resistance of plants to toxins

Rudolf Mitterbauer* and Gerhard Adam**
Centre of Applied Genetics, University of Agricultural Sciences Muthgasse, 18/05166, A-1190 Vienna, Austria;
**Present address: Carlsberg Research Laboratory, Gamle Carlsberg Vej 10, DK-2500 Valby, Denmark;*
***Author for correspondence (Fax: +43136006-6392; E-mail: adam@edv2.boku.ac.at)*

Key words: mycotoxin, deoxynivalenol, drug efflux, ribosomal protein L3, trichothecene

Abstract

Secondary metabolites produced by pathogens during the infection process are thought to play a role as pathogenicity or virulence determinants in many plant diseases. Baker's yeast and the plant *Arabidopsis thaliana* are attractive models for elucidating molecular mechanisms of resistance to toxic substances. For the *Fusarium* mycotoxin deoxynivalenol, the following resistance mechanisms were identified in yeast: (1) reduced toxin uptake due to the ABC transporter protein Pdr5p (molecular efflux pump), (2) detoxification by the acetyltransferase Ayt1p, and (3) modification of the ribosomal target by amino acid changes in the ribosomal protein L3 (Rpl3p). *PDR5*-like genes exist in plant genomes as large gene families and could play an important role as a first line of defence against a broad range of toxic metabolites. Amino acid alterations in the highly conserved *RPL3* genes could likewise play a role in trichothecene resistance in plants. The knowledge obtained using model systems should be valuable in biotechnological approaches to disease control and marker-assisted resistance breeding.

Abbreviations: ABC – ATP binding cassette; PDR – pleiotropic drug resistance.

Introduction

Plant pathogenic fungi frequently produce toxins in infected plants. These products can cause adverse health effects when high levels are present in food or animal feed. Mycotoxins are frequently phytotoxic, and for obvious reasons toxin production is suspected to contribute to the virulence of pathogens. A well characterized example is the interaction of *Cochliobolus carbonum* with its host plant maize. Resistance against the pathogen is caused by neutralization of the fungal pathogenicity factor (HC toxin) by the product of the *HM1* disease resistance gene, which encodes a specific detoxification enzyme (Pannaccione et al., 1992; Johal and Briggs, 1992). For most toxins, the situation is less clear. Often only circumstantial evidence is available (e.g. correlation of the amount of toxin produced with virulence of the pathogen). On the other hand, plant breeders are frequently confronted with a situation where quantitative differences in toxin resistance are observed between cultivars and these differences are correlated with resistance to the pathogen (Buiatti and Ingram, 1991). For most agriculturally important pathosystems, the molecular basis of toxin resistance in the respective host plants is not known.

Fusarium head blight (FHB, scab) of small-grain cereals is an agriculturally important disease (Parry et al., 1995; McMullen et al., 1997) causing severe yield losses and unacceptable contamination of the harvested grain with deoxynivalenol (DON). The mycotoxin DON is a representative of the class of trichothecenes, a highly diverse class of toxins acting as inhibitors of eukaryotic protein biosynthesis. Many 'pathogenicity-related' proteins are specifically induced in plants by pathogens and seem to play a central role in the defence response of plants (Kombrink and Somssich, 1997). Toxins with the potential to interfere with the expression of defence related proteins are therefore obvious candidates for virulence factors (affecting the kinetics or severity of disease development). Evidence that the production of trichothecenes contributes to the virulence of *Fusarium graminearum* on wheat has been obtained in laboratory and field studies using gene disruption mutants of the fungus (Proctor et al., 1995; Desjardins et al., 1996).

No complete resistance is known in the wheat breeding material, but quantitative differences exist and are inherited in a complex fashion. Toxin resistance seems to be a relevant component of *Fusarium* resistance (Grausgruber et al., 1998). Testing of a large number of wheat cultivars representing the full spectrum of available *Fusarium* resistance revealed a strong correlation between toxin resistance (determined using a seed germination assay, Lemmens et al., 1994), and field resistance, indicating that about 40% of the variation can be explained by differences in toxin resistance (Lemmens et al., unpublished data).

Our research goal is to identify molecular mechanisms responsible for differences in trichothecene resistance in plants, and for this purpose we have employed the yeast *Saccharomyces cerevisiae* and the plant *Arabidopsis thaliana* as model systems.

Results and discussion

Identification of candidate trichothecene resistance genes: Drug efflux pumps

Resistance to a toxic compound can be achieved in several ways: reduced net uptake, detoxification (and sequestration), and modification of the drug target. Using yeast it was found that basal resistance was determined by the level of expression of the ABC (ATP binding cassette) transporter protein encoded by the *PDR5* (pleiotropic drug resistance) gene (Adam and Lemmens, 1996). Mutants containing a deleted *PDR5* gene showed growth inhibition by DON, 3-acetyldeoxynivalenol (3-AcDON), nivalenol, T-2 toxin and all other trichothecenes tested, while wild-type yeast strains tolerated high levels of these mycotoxins.

PDR5 is one of nine members of the (NBF-TMS$_6$)2 subclass of ABC genes present in the yeast genome (Decottignies and Goffeau, 1997). Characteristic features of *PDR5*-like genes are the internal duplication and the presence of the ATP binding sites (NBF, nucleotide binding folds Walker A and Walker B) which are located N-terminally of the predicted six transmembrane spanning domains (TMS). This structure is shown in Figure 1A. It differs from the structure of the *MDR1* related ('P-glycoprotein' like) and vacuolar *MRP*-like proteins, which have also been identified in plants (Dudler and Hertig, 1992; Lu et al., 1997). Pdr5p and other proteins of this class are located in the plasma membrane, and most likely confer resistance by acting as 'molecular efflux pumps', removing toxic substances using the energy of ATP hydrolysis.

The finding that the ABC transporter Pdr5p confers resistance to trichothecenes in yeast could become relevant for efforts to increase *Fusarium* resistance of crop plants by biotechnology. First, *Agrobacterium*-mediated transformation has been used to introduce the yeast *PDR5* gene into tobacco. Increased resistance to DON was observed in several of the regenerated plants using a leaf disc regeneration assay. On the other hand, there is evidence that *PDR5*-like genes are present in plant genomes as large gene families. For instance, database searches revealed the existence of 15 genes with *PDR5*-like topology and high sequence similarity in the *Arabidopsis* genome (see Figure 1A). The nomenclature proposed in 1999 at the 9th International Congress on Molecular Plant–Microbe Interactions and the COST 835 meeting in Rome is to name these genes *AtPDRxy*, with *y* counting the number of *PDR5*-like genes on *Arabidopsis* chromosome number *x* (Mitterbauer et al., 2000). The sequence similarity between the yeast *PDR5* gene and the various plant homologues is, for instance, high in a protein fragment (see Figure 1B), which is characteristic of the *PDR5*-like subclass of ABC transporters (for review see: Decottignies and Goffeau, 1997). Our group currently works on the inducible over-expression of *PDR5*-like genes in *Arabidopsis* and on the expression of the *Arabidopsis* cDNAs in yeast mutants with the goal to learn more about the role and substrate specificity of these plant *PDR* genes.

ABC transporter proteins with their broad but distinct substrate specificity could have an important, currently disregarded role in plant defence, as a first line protection against toxic metabolites produced by plant pathogens. Over-expression of foreign ABC transporter protein genes in plants using the tools of plant biotechnology seems a straightforward strategy. Nevertheless, the finding that plant genomes already contain a bounty of *PDR* genes raises the question: why has natural selection (or selection by plant breeders)

A)

```
AtPDR11(11)    ---GRTVVCTIHQPSIDIFESFDELLLMKRGGQVIYAGSLGHHSQKLVEYFEAVEGVPKINDGYNPATWMLD---
AtPDR12(8)     ---GRTVVCTIHQPSIDIFEAFDELMLMKRGGQVIYAGPLGQNSHKVVEYFESFPGVSKIPEKYNPATWMLE---
AtPDR13(7)     ---GRTVVCTIHQPSIDIFEAFDELLLMKRGGHVIYSGPLGRNSHKVVEYFESFPGVPKIPEKYNPATWMLE---
AtPDR14(12)    ---GRTVVCTIHQPSIDIFEAFDELFLLKRGGEEIYVGPLGHESTHLINYFESIQGINKITEGYNPATWMLE---
AtPDR21(3)     ---GRTVVCTIHQPSIDIFEAFDELLLMKRGGQVIYGGKLGTHSQVLVDYFQGINGVPPISSGYNPATWMLE---
AtPDR22(4)     ---GRTIVCTIHQPSIDIFESFDELLFMKRGGELIYAGPLGQKSCELIKYFESIEGVQKIKPGHNPAAWMLD---
AtPDR23(5)     ---GRTIVCTIHQPSIHIFEAFDELVLLKRGGRMIYSGPLGQHSSCVIEYFQNIPGVAKIRDKYNPATWMLE---
AtPDR24(6)     ---GRTVVCTIHQPSIDIFESFDELLLMKRGGQVIYAGTLGHHSQKLVEYFEAIEGVPKIKDGYNPATWMLD---
AtPDR31(1)     ---GRTVVCTIHQPSIDIFEAFDELLLLKRGGQVIYAGPLGQNSHKIIEYFQAIHGVPKIKEKYNPATWMLE---
AtPDR32(9)     ---GRTIVCTIHQPSIDIFEAFDELVLLKRGGRMIYTGPLGQHSRHIIEYFESVPEIPKIKDNHNPATWMLD---
AtPDR33(10)    ---GRTVVCTIHQPSIDIFESFDELFLLTRGGEEIYVGPIGHHSSQLIEYFEGIRGVGKIKEGYNPATWALE---
AtPDR41        ---GRTVVCTIHQPSIDIFETFDELILMKDGGQLVYYGPLGKHSSKVIKYFESIPGVPKVQKNCNPATWMLD---
AtPDR42(2.2)   ---GRTVVCTIHQPSIDIFETFDELILMKNGGQLVYYGPPGQNSSKVIEYFESFSGLPKIQKNCNPATWILD---
AtPDR43        ---GRTVVCTIHQPSIDIFEAFDELILMKNGGKIIYYGPLGQHSSKVIEYFMRIHGVPKLKENSNPATWILD---
AtPDR44        ---GRTVVCTIHQPSIDIFEAFDELILMKNGGKIIYYGPLGQHSSKVIEYFMSIPGVPKLKENSNPATWILD---
ScPDR5         ---GQAILCTIHQPSAILMQEFDRLLFMQRGGKTVYFGDLGEGCKTMIDYFE-SHGAHKCPADANPAEWMLE---
```

B)

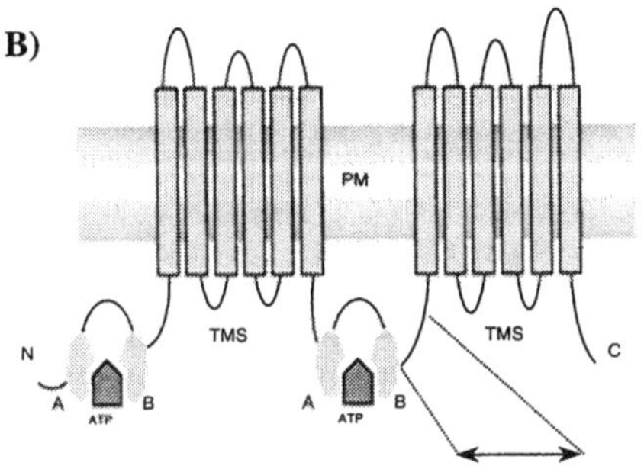

Figure 1. (A) Alignment of protein sequence fragments predicted from *Arabidopsis thaliana* genomic sequences and yeast Pdr5p. The numbers in brackets indicate the *PDR* gene numbering according to Sanchez-Fernandez et al., 2001. See note added in proof for details. (B) Model of the structure of *PDR5*-like ABC transporter proteins. Indicated are the ATP-binding folds (A, Walker A; B, Walker B) and the transmembrane spanning domains. The location of the sequence shown in (A) is indicated by the arrow.

not come up with such a simple solution? In the case of *PDR5*, over-expression leads to increased sensitivity against yet another toxic *Fusarium* metabolite. Thus, much basic research remains to be done in this area to avoid unwanted side effects in transgenic crop plants.

Detoxification

Mutants of yeast that are extremely sensitive to toxins are a valuable tool to study detoxification reactions. For instance, over-expression of the yeast *AYT1* gene, encoding a putative trichothecene 3-O-acetyltransferase, suppressed DON sensitivity of Δ*pdr5* mutants (unpublished data). The yeast *AYT1* gene is highly related to the *Tri101* gene found in *Fusarium* (Kimura et al., 1998). Detoxification of trichothecenes *in planta* could be an attractive approach. Yet, the identified acetyl-transferase, which converts DON into 3-AcDON, does not seem to be a prime candidate for a biotechnological application. The toxicity of 3-AcDON towards mammals is only slightly lower than that of DON. Furthermore, most *Fusarium* strains seem to initially produce acetylated compounds. Plant carboxylesterases, in addition to fungal enzymes, may play a significant part in the conversion of the acetylated precursors to DON. Consequently, the effectiveness of over-expression of an acetyl-transferase in plants could be severely limited by futile cycling. We are therefore studying additional candidate detoxification enzymes from yeast and other sources, which hopefully will lead to a more pronounced and irreversible detoxification.

Target modification

Trichothecene-sensitive Δ*pdr5* mutants of yeast were also employed to identify alterations of the ribosomal target of trichothecenes, which confer toxin resistance. To this end, a saturation mutagenesis of the gene encoding ribosomal protein L3 was performed. A library of mutagenized plasmids was transformed into a toxin sensitive host strain which contains solely a conditional copy of the gene encoding ribosomal protein L3 (*pGAL1-RPL3,* which is not expressed in glucose medium). Transformants resistant to deoxynivalenol were isolated and the mutations in *RPL3* characterized by DNA sequencing (Raditschnig et al., unpublished data).

Like other ribosomal protein genes, *RPL3* is extremely well conserved in evolution, and it is an attractive biotechnological approach to introduce the corresponding amino acid changes (which confer semi-dominant toxin resistance in yeast) into a plant *RPL3* homologue. On the other hand, the knowledge of the spectrum of mutations conferring resistance in yeast could also be useful for screening *Fusarium* resistant germplasm and for the development of molecular markers for plant breeding. Ribosomal resistance has, for instance, been suggested for the highly *Fusarium* resistant wheat cultivar Frontana (Miller and Ewen, 1997).

In summary, we have tried to use the wealth of 'genomics' information already available for yeast and *Arabidopsis* to gain insights into an important aspect of the problem of mycotoxins in agriculture. The identified mechanisms of trichothecene resistance – especially drug efflux pumps and mutations of the ribosomal target – are most likely relevant for crop plants. Knowledge of candidate resistance mechanisms and genes may be advantageous especially in the case of *Fusarium* diseases, where many important crop plants (wheat, durum wheat, barley, rye, corn) are affected, and when breeding for increased *Fusarium* resistance is pursued in parallel.

Acknowledgements

This work was supported by the Austrian Academy of Sciences (APART Fellowship to GA), the Austrian Science Foundation (FWF P-11551) and the City of Vienna (Hochschuljubiläumsstiftung H-239/99).

Note added in proof:

In their work on the *Arabidopsis thaliana* ABC protein superfamily Sanchez-Fernandez et al. (2001) claim that 13 *PDR5*-like ABC transporters exist, numbering the genes *PDR1* to *PDR13* (see also: http://www.arabidopsisabc.net/). According to our analysis *PDR11* and *PDR13* are the same genomic sequence from different BACs, spliced differently. We also arrive at a slightly different splicing pattern for *PDR3*, *PDR6*, *PDR7*, and *PDR10*. We have identified a cluster of 4 tandemly repeated genes on chromosome IV, only the second gene has been recognized by Sanchez-Fernandez et al. but spliced incorrectly '*PDR2.2*'. We have cloned and sequenced the full length cDNA (Mitterbauer, 2000; and unpublished data).

References

Adam G and Lemmens M (1996) Involvement of ABC transporter proteins in trichothecene resistance. Abstract B3, 8th IS-MPMI Conference, Knoxville, TN

Buiatti M and Ingram DS (1991) Phytotoxins as tools in breeding and selection of disease-resistant plants. Experientia 47: 811–819

Decottignies A and Goffeau A (1997) Complete inventory of the yeast ABC proteins. Nature Genetics 15: 137–145

Desjardins AE, Proctor RH, Bai G, McCormic SP, Shaner G, Buechly G and Hohn TM (1996) Reduced virulence of trichothecene non-producing mutants of *Gibberella zeae* in wheat field tests. Molecular Plant-Microbe Interactions 9: 775–781

Dudler R and Hertig C (1992) Structure of an mdr-like gene from *Arabidopsis thaliana*. Journal of Biological Chemistry 267: 5882–5888

Grausgruber H, Lemmens M, Bürstmayr H and Ruckenbauer P (1998) Chromosomal location of *Fusarium* head blight resistance and *in vitro* toxin tolerance in wheat using the Hobbit 'sib' (*Triticum macha*) chromosome substitution lines. Journal of Genetics and Breeding 52: 173–180

Johal GS and Briggs SP (1992) Reductase activity encoded by the HM1 disease resistance gene in maize. Science 258: 985–987

Kimura M, Kaneko I, Komiyama M, Takatsuki A, Koshino H, Yoneyama K and Yamaguchi I (1998) Trichothecene 3-O-acetyltransferase protects both the producing organism and transformed yeast from related mycotoxins. Journal of Biological Chemistry 273: 1654–1661

Kombrink E and Somssich IE (1997) Pathogenesis-related proteins and plant defense. In: Tudzynski C (ed) The Mycota V, Part A, Plant Relationships (pp 107–128). Springer Verlag, Berlin

Lemmens M, Reisinger A, Bürstmayr H and Ruckenbauer P (1994) Breeding for head blight (*Fusarium* ssp.) resistance in wheat: Development of a mycotoxin-based selection method of seedlings. Acta Horticulturae 355: 223–232

Lu YP, Li ZS and Rea PA (1997) *AtMRP1* gene of *Arabidopsis thaliana* encodes a glutathione S-conjugate pump: Isolation and functional definition of a plant ATP-binding cassette transporter gene. Proceedings of the National Academy of Sciences of the USA 94: 8243–8248

McMullen M, Jones R and Gallenberg D (1997) Scab of wheat and barley: A re-emerging disease of devastating impact. Plant Disease 81: 1340–1348

Miller JD and Ewen MA (1997) Toxic effects of deoxynivalenol on ribosomes and tissues of the spring wheat cultivars Frontana and Casavant. Natural Toxins 5: 234–237

Mitterbauer R (2000): Molecular mechanisms of trichothecene resistance and development of a yeast-based bioassay for zearalenone. Ph.D. thesis, University of Agricultural Sciences, Vienna, Austria.

Mitterbauer R, Karl T, Lemmens M, Kuchler K and Adam G (2000) Resistance to mycotoxins: A role for ABC transporter proteins in plant–pathogen interactions. In: de Wit PJGM, Bisseling T and Stiekema WJ (eds) Biology of Molecular-Plant Microbe Interactions Vol. 2 (pp 352–355) IS MPMI, St. Paul, MN, USA.

Panaccione DG, Scott-Craig JS, Pocard JA and Walton JD (1992) A cyclic peptide synthase gene required for pathogenicity of the fungus *Cochliobolus carbonum* on maize. Proceedings of the National Academy of Sciences of the USA 89: 6590–6594

Parry DW, Jenkinson P and McLeod L (1995) *Fusarium* ear blight (scab) in small grain cereals – a review. Plant Pathology 44: 207–238

Proctor RH, Hohn TM and McCormick SP (1995) Reduced virulence of *Gibberella zeae* caused by disruption of a trichothecene toxin biosynthetic gene. Molecular Plant-Microbe Interactions 8: 593–601

Sanchez-Fernandez R, Davies TGE, Coleman JOD and Rea PA (2001) The *Arabidopsis thaliana* ABC protein superfamily, a complete inventory. Journal of Biological Chemistry 276: 30231–30244.

European Journal of Plant Pathology **108:** 705–711, 2002.

Mycotoxin genetics and gene clusters

G.S. Sidhu
Department of Biology California State University, Fresno, CA 93740, USA (Fax: +15592783963)

Key words: mycotoxin, fungal genetics, *Aspergillus* spp., gene clusters, super-gene, genetics of aflatoxins

Abstract

Fungi produce low molecular weight secondary metabolites such as antibiotics and mycotoxins. Antibiotics cure diseases whereas mycotoxins cause diseases in plants, animals and human beings. Species such as *Aspergillus*, *Fusarium*, *Penicillium* and *Stachybotrys* are known to produce mycotoxins that accumulate in processed foods and feeds, although the incidence of infection occurs before processing, during the active growth of the organism. Among the mycotoxins, aflatoxins produced by *Aspergillus flavus* and *A. parasiticus* have been extensively studied at the molecular level. A complex biosynthetic pathway involving sixteen steps is mediated by individual major genes. These fungi have eight linkage groups, but the aflatoxin/sterigmatocystin (AF/ST) metabolic pathway genes have been mapped to only three linkage groups; ten of them belong to linkage group VII, and one of each to linkage group II and VIII. These genes are involved in both the regulatory and biosynthetic pathways and are clustered on the respective chromosomes. Clustering of genes in fungi indicates an evolutionary trend among genes that orchestrate gene function. Being linked together they segregate 'as a unit', thereby conferring a selective advantage to the organism. The evolution of gene clusters takes place through vertical or horizontal gene transfer. In fungi, horizontal gene transfer is most effective. Functionally, the mechanism of evolution of mycotoxin gene clusters in fungi seems to be similar to the evolution of a super-gene. The possible implications of evolutionary parallelism of gene clusters and super-genes is briefly explored.

Introduction

Many fungal species produce low molecular weight secondary metabolites. Some of them are economically important, either being useful or harmful to humans. Among the useful ones are the antibiotics, while the harmful ones include the mycotoxins. One of the most significant effects of these toxins is the induction of mycotoxicosis i.e. production of diseases in plants, animals and humans. An early record of mycotoxicosis is the occurrence of 'Turkey X' diseases in the 1960s (Lancaster et al., 1961). In cattle, it causes vomiting and sterility and in humans it induces vomiting, stomach ache and muscle pains. Usually, mycotoxicosis occurs as a result of eating crop products such as fodder, feed or food. Among the fungal genera that are known to produce mycotoxins, *Penicillium, Fusarium, Stachybotrys*, and *Aspergillus* are best known *Aspergillus, Penicillium*, and *Fusarium* often produce toxins in stored grains and hay, and in processed foods and feeds, although the incidence of infection takes place in the field. *Stachybotrys* species produce toxins by colonizing straw, hay or other cellulose rich products such as fodder and animal bedding. *Fusarium*-related toxins are primarily produced on maize and other cereal grains. *Aspergillus* species are notorious for producing aflatoxins in infected cereal grains and legumes, usually at low concentrations. However, in peanuts, cotton seed and fishmeal, they can produce high concentrations of toxins. Disease symptoms caused by aflatoxins in humans and animals are related to the type of toxins produced and to the age of the animal. Low doses of aflatoxins fed over a long period of time to pregnant sows, calves, fattening pigs, sheep, and mature cattle can led to abortion, reduced growth, nausea, feed refusal, and eventually predisposition to other serious diseases such as liver and kidney cancer. Among the mycotoxins, aflatoxins produced by *Aspergillus* are the more important because of their economic importance.

Aspergillus species

Aflatoxins are highly toxic secondary metabolites produced by *Aspergillus* species such as *A. flavus* and *A. parasiticus.* Their attributes include antibiotic activity, carcinogenicity and toxicity in humans. Betalactam can be used to treat human diseases, but aflatoxin B_1 (AFB_1) causes liver cancer. Many aflatoxins, when present in foods, cause chronic aflatoxicosis in animals leading to haemorrhaging, suppression of immunity and loss in weight-gain. The importance of aflatoxins in human and animal diseases warrants an understanding of the genetic basis of their biosynthetic pathways.

Aflatoxins are synthesized through a complex pathway involving at least sixteen steps following the synthesis of the first stable intermediate, norsolorinic acid (NOR) (Bhatnagar et al., 1992). A proposed biosynthetic pathway scheme involved in aflatoxin/sterigmatocystin (AF/ST) is shown in Figure 1. This review will summarize the genetic steps involved in the pathway with emphasis on the mechanism of clustering of genes and the possibility of evolution of a 'super-gene'.

Genetics of aflatoxin production

The genetic basis of aflatoxin production in fungi has been extensively studied in *A. flavus, A. parasiticus,* and *A. nidulans*. These fungi lack a regular sexual cycle, but have a well-defined parasexual cycle as an alternative to sexual reproduction (Papa, 1973). Briefly, the parasexual cycle involves fusion between unlike haploid nuclei leading to the formation of a heterokaryon in which crossing over occurs. Crossing over assures a reassortment of the genes on the chromosomes and eventually the chromosomes reshuffle through haploidization. The sequence of events of a parasexual cycle mimics the effects of a regular sexual cycle. Parasexuality is usually limited to intra-strain fusion. In inter-strain fusion heterokaryon incompatibility ensues and therefore inter-strain or inter-specific crosses are limited. Heterokaryon incompatibility is used as an experimental tool to establish vegetative compatibility groups (VCG) in facultative fungi.

In *A. flavus* from maize, thirty-two strains were used to generate heterokaryons by mating them in all possible combinations. Twenty-two VCGs were detected (Bennett and Papa, 1988). Fundamental genetic work of Papa and his associates identified eight linkage groups (= 8 chromosomes) in *A. flavus* using both aflatoxin and morphological auxotrophic mutants (see Bennett and Papa, 1988). Six genes controlling auxotrophy and conidial mutants have been identified in *A. flavus*. Additionally, twenty-six strains were mutated for aflatoxin biosynthesis. All auxotrophic, conidial and aflatoxin mutants were mapped to eight linkage groups. Aflatoxin genes were mapped to only three linkage groups; ten of them belong to linkage group VII, and one of each to linkage group II and linkage group VIII. The aflatoxin alleles were recessive to the dominant alleles. The only dominant allele controlling aflatoxin was due to a deletion of over 120 kb (Woloshuk et al., 1995).

Originally eight linkage groups were identified on the basis of genetic analysis of morphological and aflatoxin mutants (Bennett and Papa, 1988). However, recent studies indicate that there may only be seven chromosomes in *A. flavus* and *A. parasiticus*. Payne has suggested (Personal Communications) that the heavier band (7.0 Mb) may in fact be a composite of two separate bands having approximately the same molecular weight.

Biosynthetic pathways

The biochemistry of aflatoxin biosynthesis has been reviewed recently (Payne and Brown, 1998; Hicks et al., 2001). Briefly, there are sixteen enzymatic steps starting from a polyketide precursor to AF/ST biosynthesis (Figure 1). The first step involves a fatty acid synthase (FAS) enzyme that turns acetate units into a fatty acid precursor molecule (Brown et al., 1996a; Watanabe et al., 1996). Genetic evidence indicates that a single gene is involved in this first stage (Mahanti et al., 1996). The precursor molecule (a polyketide) is converted by polyketide synthase (PKS) to an unstable intermediate, noranthrone, which is immediately converted to a stable molecule NOR. However, the conversion of an intermediate molecule to NOR has been proposed (see Hicks et al., 2001). In *A. nidulans,* the FAS is composed of two distinctive FAS complexes, one is needed for primary fatty acid metabolism and the other for sterigmatocystin (ST) biosynthesis. Four genes are involved and two of them, *stc J* and *stc K*, are

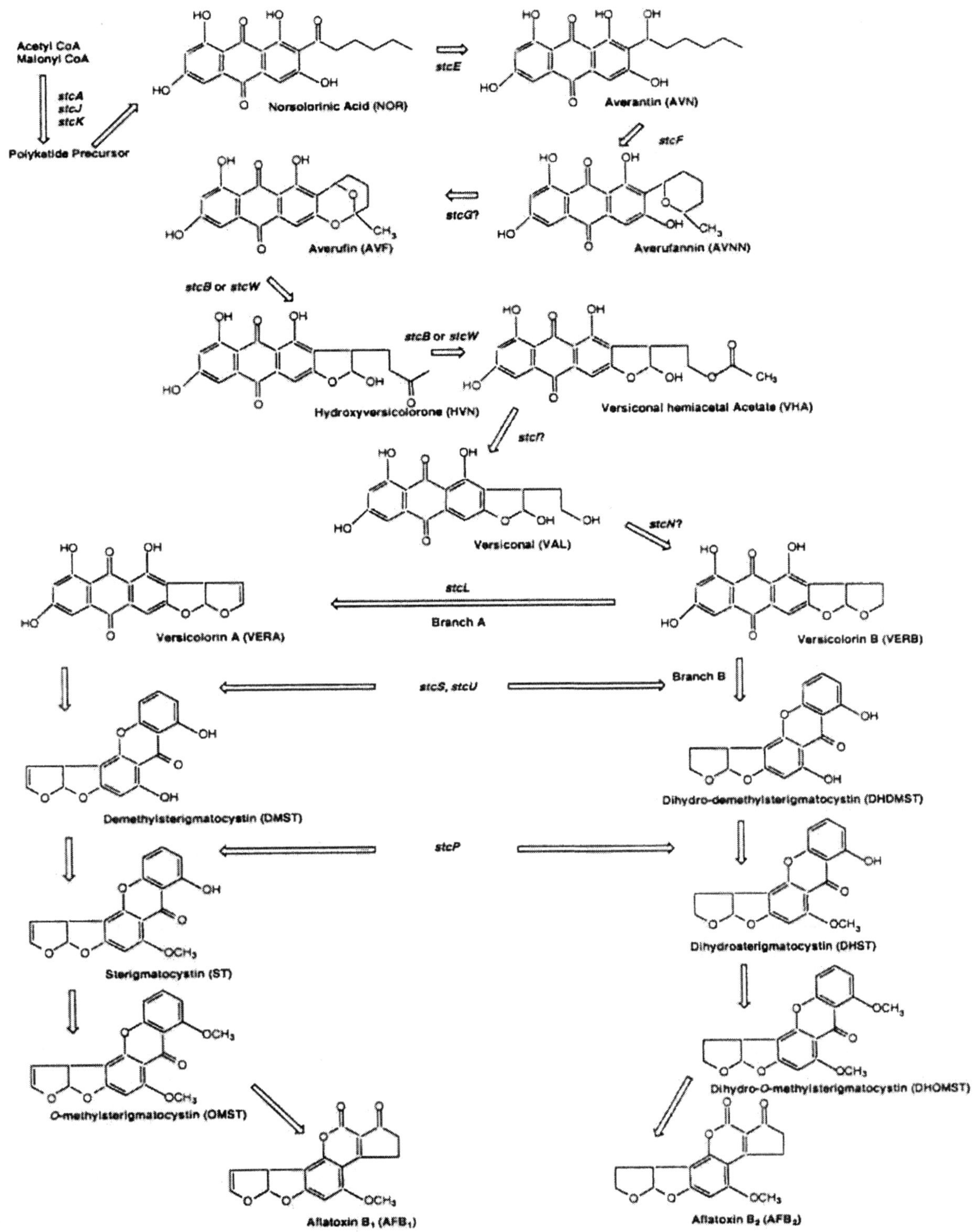

Figure 1. The AF/ST biosynthetic pathway. Malonyl CoA and acetyl CoA are converted to a polyketide precursor by the combined activities of a FAS and PKS. The polyketide precursor undergoes a series of conversions which results in formation of ST (*A. nidulans*), AFB_1/AFB_2 (*A. flavus*) or $AFB_1/AFB_2/AFG_1/AFG_2$ (*A. parasiticus*). The *A. nidulans stc* (sterigmatocystin cluster) genes responsible for producing various enzymes in the pathway are indicated. A question mark following the gene indicates that function has not been demonstrated and activity is only postulated (reproduced from Julie K. Hicks, Kiminori Shimizu and Nancy P. Keller (2001)).

part of a ST gene cluster. These genes are part of the FAS-α and FAS-β subunit domains.

The first stable compound, NOR, is converted to averantin (AVN) with the help of two enzymes, though their nature is obscure (Bhatnagar et al. 1992). In *A. parasiticus*, Chang et al. (1992) isolated a gene (*nor-1*) that complemented a NOR accumulation defect. Another gene *norA*, with a similar function, has been shown to convert NOR to AVN (Cary et al., 1996). In *A. nidulans*, a single gene (*stc E*) is involved in conversion of NOR to AVN (Brown et al., 1996a,b; Butchko et al., 1999). The AVN conversion to averufin (AVF) is carried out by at least two enzymes. The first reaction yields an intermediate compound, 5′ hydroxyaverantin (HAVN), and the second converts HAVN to AVF (Yabe et al., 1991; Yu et al., 1997; Chang et al., 2000). AVF is first converted to hydroxyversicolorone (HVN) and then to versiconal hemiacetal acetate (VHA). VHA is converted to versiconal (VAL) and VAL to versicolorin B (VERB). VERB is oxidized to versicolorin (VERA). VERA and VERB initiate pathway A and B, respectively, using the same enzymes (Figure 1). In pathway A, VERA is converted to demethylsterigmatocystin (DMST) followed by the production of sterigmatocystin (ST) in *A. nidulans* (Kelkar et al., 1996; 1997). In *A. flavus*, both pathways produce (AFB_1) and aflatoxin B_2 (AFB_2), whereas *A. parasiticus* produces AFB_1/AFG_1 and AFB_2/AFG_2 as their end products (Yabe et al., 1999).

Genes involved in pathways

Regulatory and biosynthetic genes have been categorized in the AF/ST pathway in *Aspergillus*. Both types of genes are clustered in a 60–75 kb region of the DNA (Brown et al., 1996b; Trail et al., 1995; Yu et al., 1995). Analysis of gene clusters revealed that *A. flavus, A. parasiticus* and *A. nidulans* contain the same genes in their AF/ST pathway, whereas the order and direction of ST gene expression in *A. nidulans* are rearranged (Cary et al., 2000).

Gene clusters and super-genes

A successful adaptation of an organism to its environment requires the cooperation of many genes that contribute to its survival. Genes assort independently when present on separate chromosomes or are separated by crossing over when present on the same chromosome. If the genes are tightly linked, they will rarely be broken apart by chiasmata and thus segregate together as a unit. Clustering of genes in fungi indicates tight linkage among the genes that orchestrate gene function and that segregate as a unit.

Functionally, the evolution of gene clusters in fungi and bacteria is identical to super-gene formation (Darlington and Mather, 1949). However, super-gene formation in diploid organisms has its basis in polymorphic adaptation, which in turn relies on heterozygous advantage. Heterozygous advantage is established in sexually reproducing organisms, but not in haploid organisms. However, the occurrence of the parasexual cycle in some fungi may substitute for heterozygous advantage.

The control of polymorphic adaptations is different in diploid organisms in the sense that genes must be held together so that their alleles can be segregated from one another in a block called a *super-gene*. The loci of a super-gene may become separated by rare chiasmata formation, which produces an ill-adapted group that will be eliminated like a disadvantageous mutant. If the linked genes interact usefully, structural changes to bring them on the same chromosome will be favoured.

In haploid organisms such as *Neurospora* and *Aspergillus*, polymorphism may be maintained through *frequency-dependent selection*. Frequency-dependent selection implies that an allele tend to be selected *for* when it is useful and *against* when it is not useful or is neutral to the organism. This type of selection is only plausible in haploid organisms through mutations. Useful mutations tend to accumulate on a single chromosome and this results in establishing stable coexistence of a variety of alleles in the populations. When the useful alleles form a cluster or a super-gene, heterokaryon incompatibility keeps them isolated in natural populations. In *A. nidulans,* Grindle (1963) found that independently isolated wild-type strains were often mutually incompatible. Compatibility often occurs between strains found in the same habitat. Widespread incompatibility in fungi suggests delineation of related species such as *A. flavus and A. parasiticus.* These fungi are mainly responsible for aflatoxin formation and possess a common aflatoxin pathway. The genes that control the enzymatic steps in the pathway are clustered and seem to be identical to the super-gene category of diploid organisms.

Early examples of clustering of genes are found in carbon-catabolic pathways in fungi. The utilization of galactose in yeast and quinate in *Neurospora crassa* are two examples of gene clusters in which functionally related genes in contiguous groups are subject to common transcription control resulting in the release of common messenger RNA – a situation analogous to the 'operon system'. In *Aspergillus*, the clustering of genes which have a common regulatory function seems to be identical to quinate catabolism in *Neurospora* (Case and Giles, 1977), but may not be candidates for operon formation. Clustering of genes or super-gene formation in *Aspergillus* has an adaptive advantage under adverse ecological conditions. The aflatoxins are produced under the adverse conditions which occur prior to harvest – a measure facilitating survival under dry conditions.

Evolution of gene clusters

Genes that orchestrate secondary metabolites are usually clustered on chromosomes. Antibiotic and mycotoxin-producing genes are good examples of gene clusters in bacteria and fungi. For example, production of the antibiotic B-lactam and AF/ST in *Penicillium* and *Aspergillus* are under the control of gene clusters (Gutierrez et al., 1999; Brown et al., 1996a,b). Genes for penicillin and cephalosporin are also present in bacteria. The regulation of gene clusters in bacteria is usually under the control of an 'operon' system, whereas such control is lacking in fungi (Keller and Hohn, 1997). However, the carbon-catabolic pathway of utilization of galactose in yeast and quinate in *Neurospora crassa* may conform to operon organization of genes suggesting a common transcriptional control. Structural gene clusters in mycotoxin-producing fungi do not transcribe m-RNA in one piece, therefore, do not operate as an 'operon.'

The evolution of gene clusters has been suggested to take place through horizontal gene transfer among organisms (Rosewich and Kistler, 2000). As compared to vertical gene transfer from common ancestors, horizontal gene transfer takes place among parallel and related species. Clustering of genes in mycotoxin-producing fungi seems to evolve through horizontal gene transfer. In this regard, a 'selfish operon' hypothesis has been proposed (Lawrence and Roth, 1996). This hypothesis proposes the clustering of genes through genetic rearrangements that link together genes which operate in a single metabolic pathway such as the AF/ST pathway in *Aspergillus*.

If any set of linked genes interact usefully, structural interchanges bringing them on the same chromosome will be favoured and selected during natural selection. Inversions and translocations have been suggested as the most effective chromosomal changes in bringing together genes that have an adaptive positive value. In the case of mycotoxin-producing fungi, the cooperating genes have an adaptive value for the organism enabling them to colonize ecological niches that would otherwise pose adverse environmental conditions. Mycotoxin-producing fungi experience adverse niches such as dry grains or straw where the development of the organism is under test. Such development can be best served by those species that orchestrate their genes in a clustered form.

The AF/ST pathway shows common genes among *A. parasiticus, A. flavus*, and *A. nidulans*. These genes are clustered in all three species (Payne and Brown, 1998). Aflatoxin genes, which seem to have common ancestral origins, accumulated in these species through horizontal transfer. The gene clusters in these fungi are still evolving because introgressed DNA present in a 60–75 kb chromosomal region contains loosely linked genes. Such pieces of introgressed DNA are usually junk DNA and have no purpose. Junk DNA would be lost over time through deletions and the toxin genes would thus become more closely linked.

Horizontal gene transfer has also been demonstrated in the case of penicillin biosynthetic gene clusters in *P. chrysogenum*. DNA sequences of several naturally occurring mutants of *P. chrysogenum* show a single copy of the gene cluster, as well as multiple copies of biosynthetic gene clusters. The borders of a single copy or multiple copies of gene clusters are flanked by an identical hexanucleotide repeat (TTTACA). Single copy cluster or multiple copies of clusters is evident from the amount of penicillin produced by the respective strains. The occurrence of the hexanucleotide repeat border, which occurs in many penicillin-producing strains, suggests a common origin and transfer of gene clusters in Penicillium. Similar mechanisms seem to hold true in *Aspergillus*.

Function of gene clusters in *Aspergillus* species

Biosynthetic genes

Sixteen genes have been identified in the AF/ST biosynthetic pathway of *A. flavus and A. parasiticus* (see Hicks et al., 2001). These genes are clustered on a 60–75 kb section of DNA (Brown et al., 1996a,b). The gene clusters of the two species contain almost the same genes (Cary et al., 2000; Keller and Hohn, 1997). However, the gene order and directions of their expression is different from *A. nidulans*. This may have occurred because *A. nidulans* has a different developmental life cycle. The gene clusters of the *Aspergillus* have a common regulatory control.

The aflR *gene*

The *aflR* (aflatoxin regulation) gene is essential for positive co-regulation of AF/ST gene clusters (Payne and Brown, 1998). This observation is supported by the fact that a deletion in or a mutation of the *aflR* gene in *A. flavus* and *A. parasiticus* impaired the AF biosynthetic ability of the biosynthetic genes. When the impaired regulatory gene was complemented with its wild-type allele, the function of the biosynthetic genes was regained (Chang et al., 1993; Yu et al., 1996a). Furthermore, the number of copies of the *aflR* gene in three *Aspergillus* species directly influenced the corresponding level of transcription of the AF/ST genes (Chang et al., 1995a).

The *aflR* genes from three species produced functionally homologous proteins, however, their DNA sequences varied greatly (Yu et al., 1996). Payne and Brown (1998) showed that *A. flavus* and *A. parasiticus* share 95% DNA sequence homology while *A. nidulans* only share 33% identity with the other two species. The *aflR* genes produce proteins that belong to a zinc-binuclear cluster (Zn(II)2 Cys6) family (Chang et al., 1993; Todd and Adrianopoulus, 1997). The base, encoding the zinc-binuclear cluster binding domain showed 71% identity with *aflR* genes of *A. flavus* and *A. nidulans*. The protein from the zinc-binuclear cluster domain binds to and recognizes a palindromic sequence in the promoters of biosynthetic genes. The first primary binding sequence, 5′-TCGN(5) CGA-3′, of *aflR* was identified in *A. nidulans* (Fernandes et al., 1998) and the second *aflR* binding sequence, 5′-TTAGGCCTAA-3′, was reported from *A. flavus* and *A. parasiticus* (Chang et al., 1995b; Payne and Brown, 1998). However, the first primary sequence remains the major binding site in *A. parasiticus* (Ehrlich et al., 1999). Interestingly, the primary sequence is conserved in the promoters of AF genes while it is repeated one or more times in the promoters of the ST cluster. Other aspects of the *aflR* gene are discussed by Hick et al. (2001). Importantly, *aflR* is not sufficient to activate AF/ST biosynthetic genes, particularly during primary growth. Several reports suggest that inhibitory factors may be operative during the early periods of fungal growth (Butchko et al., 1999). The data indicate that there may be post-transcriptional or post-translational regulation of aflR by other proteins (see Hicks et al., 2001).

References

Bennett JW and Papa KE (1988) The aflatoxigenic *Aspergillus* Spp. In: Sidhu GS (ed.) Advances in Plant Pathology, Vol 6 Genetics of Plant Pathogenic Fungi, (pp 263–280) Academic Press, London,

Bhatnagar D, Ehrlich KC and Cleveland TE (1992) Oxidation–reduction reactions in biosynthesis of secondary metabolites. In: Bhatnagar D, Lillehoj EB and Arora DK (eds) Handbook of Applied Mycology: Mycotoxins in Ecological Systems, Marcel Dekker, New York, pp. 225–286

Brown DW, Yu J-H, Kelkar HS, Fernandes M, Nesbitt TC, Keller NP, Adams TH and Leonard TJ (1996a) Twenty-five coregulated transcripts define a sterigmatocystin gene cluster in *Aspergillus nidulans*. Proceedings of the National Academy Sciences USA 93: 1418–1422

Brown DW, Adams TH and Keller NP (1996b) *Aspergillus* has distinct fatty acid synthases for primary and secondary metabolism. Proceedings of the National Academy Sciences USA 93: 14873–14877

Butchko RAE, Adams TH and Kellar NP (1999) *Aspergillus nidulans* mutants defective in *stc* gene cluster regulation. Genetics 153: 715–720

Cary JW, Wright M, Bhatnagar D, Lee R and Chu FS (1996) Molecular characterization of an *Aspergillus parasiticus* dehydrogenase gene, *nor A*, located on the aflatoxin biosynthesis gene cluster. Applied and Environmental Microbiology 63: 360–366

Cary JW, Chang PK and Bhatnagar D (2000) Clustered metabolic pathway genes in filamentous fungi. In: Arora DK and Khachatourians G (eds) Applied Mycology and Biotechnology, Vol 1, Elsevier Science, Amsterdam

Case ME and Giles NH (1977) Gene order in the *qa* gene cluster of *Neurospore crassa.* Molecular General Genetics 147: 79–82

Chang P-K, Story CD and Linz J E (1992) Cloning of a gene associated with aflatoxin B_1 biosynthesis in *Aspergillus parasiticus*. Current Genetics 21: 231–233

Chang P-K, Cary JW, Bhatnagar D, Cleveland TE, Bennet JW, Linz JE, Woloshuk CP and Payne GA (1993) Cloning of the *Aspergillus parasiticus apa*-2 gene associated with the regulation of aflatoxin biosynthesis. Applied and Environmental Microbiology 59: 3273–3279

Chang P-K, Cary JW, Yu JH, Bhatnagar D and Cleveland TE (1995a) The *Aspergillus parasiticus* polyketide synthase gene, *pksA*, a homolog *of Aspergillus nidulans wA*, is required for aflatoxin B_1. Molecular and General Genetics 248: 270–277

Chang P-K, Ehrlich KC, Yu JH, Bhatnagar D and Cleveland TE (1995b) Increased expression of *Aspergillus parasiticus aflR*, encoding a sequence-specific DNA-binding protein, relieves nitrate inhibition of aflatoxin biosynthesis. Applied and Environmental Microbiology 61: 2372–2377

Chang P-K, Yu JH, Ehrlich KC, Boue SM, Montalbano BG, Bhatnagar D and Cleveland TE (2000) *AdhA* in *Aspergillus parasiticus* is involved in conversoin of 5′ hydroxyaveratin to averufin. Applied and Environmental Microbiology 66: 4715–4719

Darlington CD and Mather K (1949) The Elements of Genetics. Allan and Unwin, London

Ehrlich KC, Montalbano BG and Cary JW (1999) Binding of the C_6-Zinc cluster protein, *aflR*, to the promoters of aflatoxin pathway biosynthesis genes in *Aspergillus parasiticus*. Gene 230: 249–257

Fernandes M, Keller NP and Adams TH (1998). Sequence-specific binding by *Aspergillus nidulans aflR*, a C_6-Zinc cluster protein regulating mycotoxin biosynthesis. Molecular Microbiology 28: 1355–1365

Grindle M (1963) Heterokaryon compatibility of closely-related strains in the *Aspergillus nidulans* group. Heredity 18: 397–405

Gutierrez S, Casqueiro F and Martin J (1999) Gene organization and plasticity of the B-lactam genes in different filamentous fungi. Antonie Van Leeuwenhoek Journal of Microbiology and Serology 75: 8194

Hicks JK, Shimizu S and Keller NP (2001) Genetics and biosynthesis of aflatoxin and sterigmatocystin. Mycota (in press)

Kelkar HS, Keller NP and Adams TH (1996) *Aspergillus nidulans stcP* encodes an O_methyltransferase that is required for sterigmatocystin biosynthesis. Applied and Environmental Microbiology 62: 4296–4298

Kelkar HS, Skloss TW, Haw JF, Keller NP and Adams TH (1997) *Aspergillus stcJ* encodes a putative cytochrome P-450 required for bufuran desaturation during aflatoxin/sterigmatocystin biosynthesis. Journal of Biological Chemistry 272: 1589–1594

Keller NP and Hohn TM (1997) Metabolic pathway gene clusters in filamentous fungi. Fungal Genetics and Biology 21: 17–29

Lancaster MD, Jenkins FP and Phillip JM (1961) Toxicity associated with certain samples of groundnuts. Nature 192: 1095–1096

Lawrence JG and Roth JR (1996) Selfish operons: horizontal transfer may drive the evolution of gene clusters. Genetics 143: 1843–1860

Mahanti N, Bhatnagar D, Cary JW, Joubran J and Linz JE (1996) Structure and function of *fas_1A* gene encoding a putative fatty acid synthetase directly involved in aflatoxin biosynthesis in *Aspergillus parasiticus*. Applied and Environmental Microbiology 62: 191–195

Papa EE and (1973) The parasexual cycle in *Aspergillus flavus*. Mycologia 65: 1201–1205.

Payne GA and Brown MP (1998) Genetics and physiology of aflatoxin biosynthesis. Annual Review of Phytopathology 36: 329–362

Rosewich VL and Kistler HC (2000) Role of horizontal gene transfer in the evolution of fungi. Annual Review of Phytopathology 38: 325–363

Todd RB and Adrianopoulos A (1997) Evolution of a fungal regulatory gene family: The Zn(II) 2Cys6 binuclear cluster DNA binding motif. Fungal Genetics and Biology 21: 388–605

Trail F, Mahanti N, Rarck M, Mehigh R, Liang SH, Zhou R and Linz JE (1995) Physical and transcriptional map of a aflatoxin gene cluster in *Aspergillus parasiticus* and functional disruption of a gene involved early in the aflatoxin pathway. Applied and Environmental Microbiology 61: 2665–2673

Watanabe CH, Wilson D, Linz JE and Townsend CA (1996) Demonstration of the catalytic roles and evidence for the physical association of type I fatty acid synthases and a polyketide synthase in the biosynthesis of aflatoxin B_1. Chemical Biology 3: 463–469

Woloshuk CP, Yousibova GL, Rollins JA, Bhatnagar D and Payne GA (1995) Molecular characterization of the *afl-1* locus in *Aspergillus flavus*. Applied and Environmental Microbiology 61: 3019–3023

Yabe K, Nakamura M, Nakajima H, Ando Y and Hamasaki T (1991) Enzymatic conversion of norsolorinic acid to averufin in aflatoxin biosynthesis. Applied and Environmental Microbiology 57: 1340–1345

Yabe K, Nakamura M and Hamasaki T (1999) Enzymatic formation of G-group aflatoxin and biosynthetic relationship between G-and B-group aflatoxin. Applied and Environmental Microbiology 65: 3807–3872

Yu J-H, Chang P-K, Cary JW, Wright M, Batnagar D, Cleveland TE, Payne GA and Linz JE (1995) Comparative maping of aflatoxin pathway gene clusters in *Aspergillus parasiticus* and *Aspergillus flavus*. Applied and Environmental Microbiology 61: 2365–2371

Yu J-H, Butchko RAE, Fernandes M, Keller NP, Leonard TJ and Adams TH (1996) Conservation of structure and function of the aflatoxin regulatory gene *aflR* from *Aspergillus nidulans* and *A. flavus*. Current Genetics 29: 549–555

Yu J-H, Chang P-K, Cary JW, Bhatnagar D and Cleveland TE (1997) AvnA, a gene encoding a cytochrome P_450 monooxygenase, is involved in the conversion of averantin to averufin in aflatoxin biosynthesis in *Aspergillus parasiticus*. Applied and Environmental Microbiology 63: 1349–1356

European Journal of Plant Pathology **108:** 713–718, 2002.

Biosynthesis of depsipeptide mycotoxins in *Fusarium*

Till Hornbogen, Mirko Glinski and Rainer Zocher*
*Fakultät II, Institut für Chemie, Arbeitsgruppe Biochemie und Molekulare Biologie, Technische Universtät Berlin, Franklinstrasse 29, D-10587 Berlin, Germany; *Author for correspondence (Phone: +493031424168; Fax: +493031473522; E-mail: razzo@chem.tu-berlin.de)*

Key words: enniatin synthetase, *Fusarium*, *N*-methylated cyclodepsipeptides, nonribosomal peptide synthesis, phytopathogenic

Abstract

The cyclic hexadepsipeptide enniatin is known as a phytopathogenic compound from Fusaria causing necrosis and wilt. The molecule consists of three alternating residues each of a branched chain amino acid and D-hydroxyisovaleric acid (D-Hiv). Enniatins are synthesized by a 347 kDa multienzyme (enniatin synthetase) via a thiol template mechanism. The corresponding gene *esyn1* has an open reading frame of 9393 nucleotides and harbours two modules, one responsible for D-hydroxy acid activation and one for L-amino acid activation with an integrated *N*-methyltransferase domain. Such methyltransferases build an homologous group among *N*-methyl peptide synthetases. Enniatins are synthesized by step-wise condensation of dipeptidol building blocks in an iterative manner resembling fatty acid synthesis. A key enzyme in enniatin biosynthesis is the NADPH-dependent D-2-hydroxyisovalerate dehydrogenase, that supplies enniatin synthetase with D-Hiv. Enniatins contribute to the wilt toxic character of Fusaria. Virulence was significantly reduced in *F. avenaceum* after disruption of the *esyn1* gene.

Introduction

Fungi constitute a diverse and damaging group of plant pathogens and therefore have a major economic importance in agriculture (Knogge, 1996). Fungi of the genus *Fusarium* are widespread phytopathogens and produce a number of highly toxic compounds e.g. enniatin, trichothecene and the homologous T-toxin. However, relatively little is known about the possible involvement of these toxins in plant infection. Most of the mycotoxins produced by *Fusarium* are not host-specific and affect a wide range of plants, causing symptoms such as necrotic lesions, rots and wilts (Drysdale, 1982). The toxicity of enniatins presumably derives from their ability to increase the permeability of membranes to ions and from their action as uncouplers of oxidative phosphorylation (Shemyakin et al., 1969).

The biosynthesis of enniatins is accomplished nonribosomally by the multifunctional enzyme enniatin synthetase (Zocher and Keller, 1997; Glinski et al., 2001). The enzyme is a *N*-methylcyclopeptide synthetase and represents a hybrid system of a peptide synthetase and an integrated *N*-methyltransferase (Haese et al., 1993). Peptide synthetases are organized into coordinated groups of active sites termed modules, each encompassing a length of approximately 600 amino acids corresponding to a molecular mass equivalent of ~70 kDa. Each module is responsible for catalysis of one cycle of polypeptide chain elongation and associated functional group modifications (Zocher and Keller, 1997; Glinski et al., 2001). The modules show considerable sequence conservation to each other and the number of these units equals the number of the amino acids serving as substrates of the corresponding enzymes.

In this paper, we describe the biosynthesis of the enniatin and its role in plant pathogenicity.

Mechanism of enniatin biosynthesis

Enniatins belong to the class of *N*-methylated cyclodepsipeptides, found in various strains of the genus *Fusarium* (Plattner et al., 1948). They consist

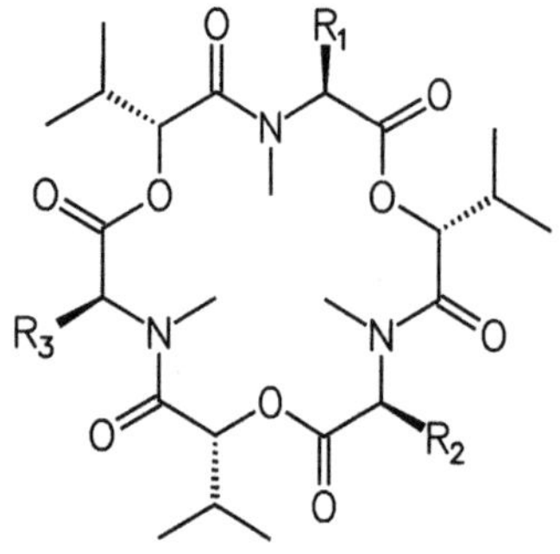

Figure 1. Structures of enniatins and beauvericin: Enniatin A: $R_1 = R_2 = R_3 =$ sec-butyl; Enniatin A1: $R_1 =$ isopropyl, $R_2 = R_3 =$ sec-butyl; Enniatin B: $R_1 = R_2 = R_3 =$ isopropyl; Enniatin B1: $R_1 = R_2 =$ isopropyl, $R_3 =$ sec-butyl; Beauvericin: $R_1 = R_2 = R_3 =$ benzyl.

of alternating residues of D-2-hydroxyisovaleric acid (D-Hiv) and a *N*-methylated branched chain L-amino acid, such as valine, leucine and isoleucine, linked by peptide and ester bonds (Figure 1). In the case of beauvericin, produced by entomopathogens, the branched chain L-amino acid is substituted by phenylalanine (Figure 1) (Hamill et al., 1969).

The biosynthesis of enniatins proceeds nonribosomally via the thiol template mechanism catalysed by the multifunctional enzyme enniatin synthetase in an iteratant process (Figure 2) (Zocher and Keller, 1997; Glinski et al., 2001). The substrates are activated as acyladenylates by the corresponding modules. Subsequently, they are covalently attached to enzyme-bound 4′-phosphopantetheine residues as thioesters. The *N*-methylation step occurs prior to peptide bond formation (Billich and Zocher, 1987). In this process, S-adenosyl-L-methionine (AdoMet) acts as the methyl group donor. Enniatin synthetase is a two-module enzyme (D-Hiv and L-amino acid activation module) (Haese et al., 1993; 1994). By interaction of the two modules a dipeptidol unit is formed, which is transferred to a thiol-waiting position (Figure 2A). This thiol group acts as the acceptor of the growing peptide chain (condensation of dipeptidol units) and picks up the intermediates of enniatin synthesis. The enniatin molecule is synthesized by three successive condensations of enzyme-bound dipeptidols. The process of enniatin formation is terminated by the cyclization reaction of the linear hexadepsipeptide (Figure 2B). Enniatin synthetases from *F. scirpi*, *F. lateritium* and *F. sambucinum* differ in their amino acid specificity, which results in the formation of various enniatins (Figure 1) (Pieper et al., 1992). This may be caused by mutations in the amino acid recognition sites of the various enniatin synthetases. The amino acid sequences of the amino acid activation domains of Esyns from *F. scirpi* and activating L-valine, and *F. sambucinum,* preferably activating L-isoleucine, are nearly identical with the exception of three regions showing significant differences (Doller et al., 1996).

Molecular structure and function of enniatin synthetase

The enniatin synthetase corresponding gene, *esyn1,* from *F. scirpi* encoding an ORF of 9393 bp (3131 amino acids) has been isolated and sequenced (Haese et al., 1993). The multifunctional enzyme is a single polypeptide of 347 kDa. Sequence analysis of the *esyn1* gene, biochemical characterization and expression studies indicated that the enzyme consists of two modules (**EA** and **EB**), which are each approximately 420 amino acid residues of size (Figure 3) (Haese et al., 1993; 1994; Pieper et al., 1992).

The N-terminal module **EA** is responsible for the activation and binding of D-Hiv. The second module **EB**, C-terminal to **EA**, activates and preferably binds L-valine. The two modules **EA** and **EB** contain a conserved 4′-phosphopantetheine binding site at the C-terminus with a highly conserved serine residue. Module **EB** has an additional 4′-phosphopantetheine group, the so-called waiting position (Figure 2). Between both modules a putative condensation (**C**) domain can be found such as in the N-terminal and C-terminal part of enniatin synthetase (Figure 3) (Glinski et al., 2001).

Module **EB** contains a *N*-methyltransferase domain **M** of 434 amino acids length (Figure 3). Domain **M** is highly conserved among *N*-methyl peptide synthetases from prokaryotic and eukaryotic origin and shows only local sequence similarities to structural elements of other AdoMet-dependent methyltransferases (motif I, II/Y, IV and V in Figure 3) (Burmester et al., 1995; Billich and Zocher, 1987; Hacker et al., 2000).

In Figure 4, an unrooted phylogenetic tree of *N*-methyltransferase domains of various peptide synthetases is shown (Glinski et al, 2001; Hacker et al., 2000). Sequence comparison revealed that the seven *N*-methyltransferase domains of the multifunctional enzyme cyclosporin synthetase, the *N*-methyltransferase domain from PF1022A synthetase (*Mycelia sterilia*) (Futamara et al., 2001)

(A)

Cycle I

P1-D-Hiv+ **P2**-MeVal ⟶ **P1**-SH +**P2**-MeVal-D-Hiv
P2-MeVal-D-Hiv+ **P3**-SH ⟶ **P2**-SH +**P3**-MeVal-D-Hiv

Cycle II

P1-D-Hiv+ **P2**-MeVal ⟶ **P1**-SH + **P2**-MeVal-D-Hiv
P2-MeVal-D-Hiv+ **P3**-MeVal-D-Hiv ⟶
P2-(MeVal-D-Hiv)$_2$ + **P3**-SH ⟶ **P2**-SH +**P3**-(MeVal-D-Hiv)$_2$

Cycle III

P1-D-Hiv+ **P2**-MeVal ⟶ **P1**-SH + **P2**-MeVal-D-Hiv
P2-(MeVal-D-Hiv) + **P3**-(MeVal-D-Hiv)$_2$ ⟶ **P2**-(MeVal-D-Hiv)$_3$+ **P3**-SH
P2-(MeVal-D-Hiv)$_3$+ **P3**-SH ⟶ **P2**-SH+ **P3**-(MeVal-D-Hiv)$_3$

Cyclization **P3**$_3$-(MeVal-D-Hiv)$_3$ **P3**-SH + **enniatin B**

(B)

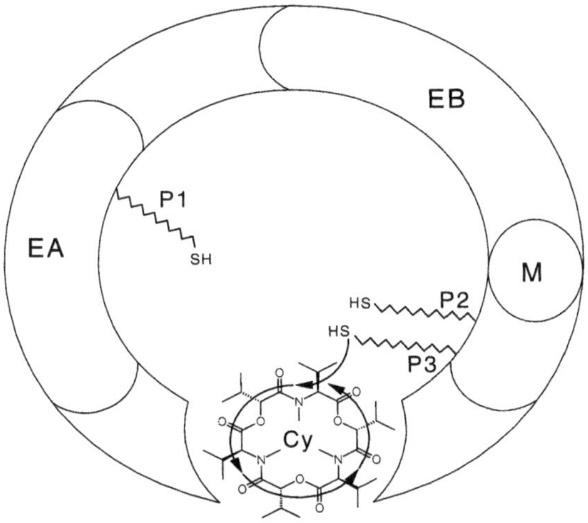

Figure 2. Mechanism of enniatin B biosynthesis. (A) Scheme of partial reactions leading to enniatin B. P1, P2, P3 = 4′-phosphopantetheine. (B) Model of arrangement of catalytic sites of enniatin synthetase. Cy: cyclization cavity; EA: D-Hiv-activation module; EB: L-valine-activation module; M: *N*-methyltransferase domain.

and the *N*-methyltransferase domains of Esyns from *F. scirpi*, *F. sambucinum* and *F. pallidoroseum* build one group (Burmester et al., 1995). The domains of the different *Fusaria* share 65% identity. The second group (prokaryotic origin) includes the *N*-methyltransferase domains from various *Streptomyces* and *Microcystis spp*. The overall sequence similarity between the two major groups is in the range of 18–25%.

Mutational analysis of the *N*-methyltransferase domain of enniatin synthetase

Figure 3 shows the sequential order of conserved motifs within the *N*-methyltransferase domain of enniatin synthetase. Deletion of the first 21 N-terminal amino acid residues of the *N*-methyltransferase domain of enniatin synthetase had no influence on AdoMet-binding in a photolabelling test with ^{14}C-AdoMet (Hacker et al., 2000). Further deletions of additional 26 amino acids resulted in a total loss of binding activity. Truncation of a short portion (38 amino acids) from the C-terminus and also deletions of internal sequences containing the conserved motifs led to complete loss of AdoMet-binding activity.

Site-directed mutagenesis of the conserved ^{2106}Tyr in motif II/Y into valine, alanine and serine significantly reduced the AdoMet-binding activity of the domain. Replacement of ^{2106}Tyr by phenylalanine restored AdoMet-binding activity to about 70%,

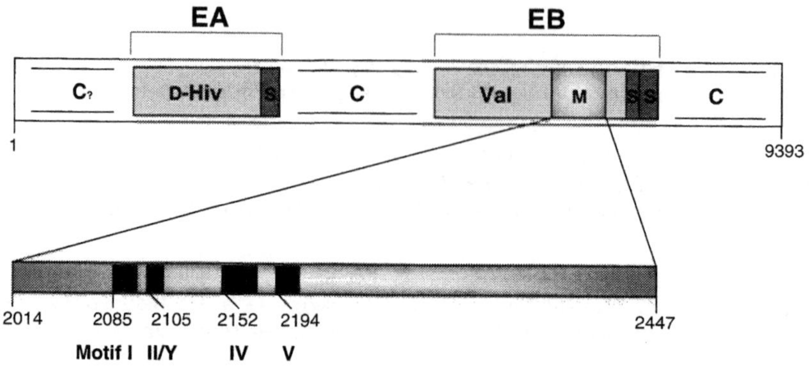

Figure 3. Physical map of enniatin synthetase (esyn1). EA represents the L-valine-activating module; EB represents the D-Hiv-activating module; S indicates the 4′-phosphopantetheine binding sites. M stands for *N*-methyltransferase domain. Putative condensation domains C are represented by an open white box. The numbers indicate the amino acid position in the sequence of enniatin synthetase. The black boxes indicate conserved motifs which can be found within various methyltransferases and *N*-methyltransferase domains of peptide synthetases.

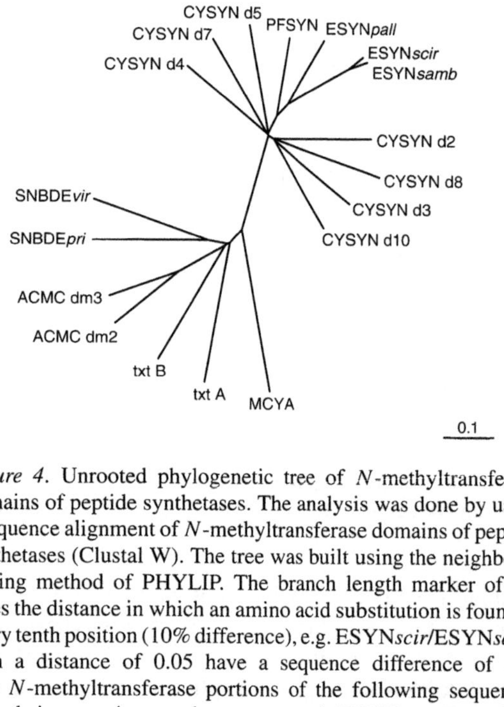

Figure 4. Unrooted phylogenetic tree of *N*-methyltransferase domains of peptide synthetases. The analysis was done by using a sequence alignment of *N*-methyltransferase domains of peptide synthetases (Clustal W). The tree was built using the neighbour-joining method of PHYLIP. The branch length marker of 0.1 gives the distance in which an amino acid substitution is found in every tenth position (10% difference), e.g. ESYN*scir*/ESYN*samb* with a distance of 0.05 have a sequence difference of 5%. The *N*-methyltransferase portions of the following sequences with their accession numbers were used: ESYN*scir*: Esyn from *Fusarium scirpi* Z18755, ESYN*samb*: Esyn from *F. sambucinum* Z48743, ESYNpall: Esyn from *F. pallidoroseum* AJ345016, CYSYN: Cyclosporin synthetase from *Tolypocladium niveum* Z28383, PFSYN: PF1022A synthetase from *Mycelia sterilia* (Futamura et al., 2001); MCYA: Microcystin synthetase A from *Microcystis spp.* AB019578, ACMC: Actinomycin synthetase II from *Streptomyces crysomallus*, SNBDE*pri*: Pristinamycin synthetase from *S. pristinaespiralis* Y11548 and SNBDE*vir*: Virginiamycin synthetase from *S. virginiae* Y11547. txt A: thaxtomin synthetase A, txt B: thaxtomin synthetase B from *S. acidiscabies* AF255732.

indicating that the tyrosine residue plays a crucial role in AdoMet-binding and that the aromatic residue of ^{2106}Tyr may be critical for AdoMet-binding in *N*-methyl peptide synthetases.

D-2-Hydroxyisovalerate dehydrogenase

In some fungal species, namely the enniatin producers of the genus *Fusarium*, there is a pathway leading from the primary precursor L-valine to D-2-hydroxyisovaleric acid via 2-ketoisovalerate (Lee et al., 1992; Zocher et al., 2000). D-Hiv dehydrogenase catalyses the reversible reaction of 2-ketoisovalerate to D-Hiv, which is an intermediate in the biosynthetic pathway of enniatins in *Fusarium*. D-Hiv dehydrogenase consists of one polypeptide chain with a molecular mass of about 53 kDa. It is strictly dependent on NADPH and exhibits a high substrate specificity with respect to 2-ketoisovalerate. This may explain the fact that D-Hiv is the exclusive hydroxy acid component in enniatins isolated from *Fusaria* (Lee et al., 1992).

Phytotoxic properties of enniatin-producing *Fusaria*

In contrast to the regulation of other secondary metabolites, the expression of the enniatin synthetase during fermentative growth follows a constitutive manner (Billich et al., 1988). To elucidate the role of enniatins in plant virulence, 36 enniatin-producing *Fusarium*

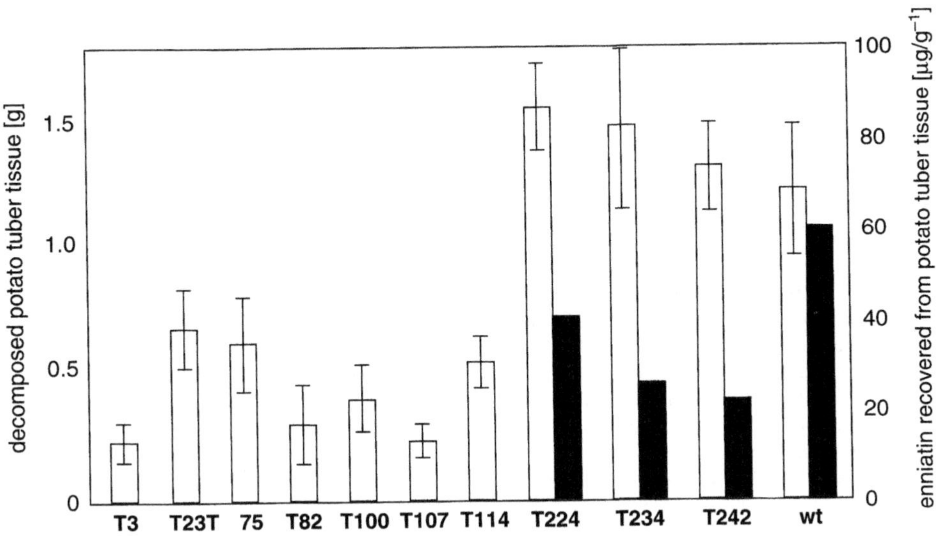

Figure 5. Virulence assay of *Fusarium avenaceum* on potato tuber tissue. Mean amounts of decomposed potato tuber tissue per slice (white bars with standard deviations), from 24 single values and means of enniatin recovered from the decomposed tissue (black bars). Values were obtained from four independent tests with six slices per test for all mutants and the wild-type (wt) strain.

strains were characterized regarding enniatin production and virulence on potato tuber tissue. Interestingly, seven enniatin-producing and 16 non-producing strains induced necrosis in the test, so that enniatin synthesis seemed not to be essential for the infection of potato tuber tissue (Herrmann et al., 1996a).

Fusarium avenaceum, a plant pathogenic strain and enniatin B producer, causes necrosis on knapweed (*Centaurea maculosa*) and also on potato tuber tissue. After disruption of the *esyn1* corresponding gene from *F. avenaceum* by homologous recombination, no enniatin production could be detected in the resulting transformants T3, T23, T75, T82, T100, T107 and T114 (Figure 5) (Herrmann et al., 1996b). The virulence of these enniatin non-producing transformants in the potato tuber tissue test was significantly reduced compared with the virulence of the group of enniatin-producing mutants (T224, T234 and T242) and the wild-type strain (Figure 5). Enniatins could be recovered from the tissue decomposed by the ectopic transformants T224, T234 and T242 and the wild-type. No enniatin was detected in tissues infected with the null mutants (Figure 5). In contrast to the findings with knapweed and potato, these results indicate a specific role of enniatin in plant pathogenesis. Further studies on other hosts, e.g. wheat, need to be carried out to obtain more detailed information on phytopathogenic mechanisms of enniatin-producing *Fusarium* strains.

References

Billich A and Zocher R (1987) *N*-Methyltransferase function of the multifunctional enzyme enniatin synthetase. Biochemistry 26: 8417–8423

Billich A and Zocher R (1988) Constitutive expression of enniatin synthetase during fermentative growth of *Fusarium scirpi*. Applied and Environmental Microbiology 54: 2504–2509

Burmester J, Haese A and Zocher R (1995) Highly conserved *N*-methyltransferase as an integral part of peptide synthetases. Biochemistry and Molecular Biology International 37: 201–207

Doller A, Haese A and Zocher R (1996) Molecular cloning of the amino acid activation domain of enniatin synthetase from *Fusarium sambucinum*. Symposium Enzymology of Biosynthesis of Natural Products. Abstract 66, 22–25 September, Technische Universität Berlin

Drysdale RB (1982) The production and significance in phytopathology of toxins produced by species of *Fusarium*. In: Moss MO and Smith JE (eds) Applied Mycology of Fungi (pp 95–105) Cambridge University Press, Cambridge, England

Futamura T, Aihara S, Midoh H, Okakura K, Kleinkauf H, Miymoto K, Watanabe M, Yanai K, Murakami T and Yasutake T (2001) Cyclic depsipeptide synthases, genes thereof and mass production system of cyclic depsipeptide. Patent AU68741000 (Meiji Seika Kaisha, Ltd.)

Glinski M, Hornbogen T and Zocher R (2001) Enzymatic synthesis of fungal *N*-methylated cyclopeptides and depsipeptides. In: Kirst H, Yeh W-K and Zmijewski M (eds) Enzyme Technologies for Pharmaceutical and Biotechnological Applications (pp 471–497) Marcel Dekker Inc., New York

Hacker C, Glinski M, Doller A, Hornbogen T and Zocher R (2000) Mutational analysis of the *N*-methyltransferase domain of the multifunctional enzyme enniatin synthetase. Journal of Biological Chemistry 275: 30826–30832

Haese A, Schubert M, Hermann M and Zocher, R (1993) Molecular characterization of the enniatin synthetase gene encoding a multifunctional enzyme catalysing *N*-methyldepsipeptide formation in *Fusarium scirpi*. Molecular Microbiology 7: 905–914

Haese A, Pieper R, von Ostrowski T and Zocher R (1994) Bacterial expression of catalytically active fragments of the multifunctional enzyme enniatin synthetase. Journal of Molecular Biology 247: 116–122

Hamill RL, Higgens CE, Boaz HE and Gorman M (1969) The structure of beauvericin: a new depsipeptide antibiotic toxic to *Artemia salina*. Tetrahedron Letters 49: 4255–4258

Herrmann M, Zocher R and Haese A (1996a) Enniatin production by *Fusarium* strains and its effect on potato tuber tissue. Applied and Environmental Microbiology 62: 393–398

Herrmann M, Haese A and Zocher R (1996b) Effect of disruption of the enniatin synthetase gene on the virulence of *Fusarium avenaceum*. Molecular Plant–Microbe Interactions 9: 226–232

Knogge W (1996) Fungal infections of plants. The Plant Cell 8: 1711–1722

Lee C, Görisch H, Kleinkauf H and Zocher R (1992) A highly specific D-hydroxyisovalerate dehydrogenase from the enniatin producer *Fusarium sambucinum*. Journal of Biological Chemistry 267: 11741–11744

Plattner PA, Nager U and Boller A (1948) Welkstoffe und Antibiotika. Siebte Mitteilung über die Isolierung neuartiger Antibiotika aus *Fusarien*. Helvetica Chimica Acta 31: 594–602

Pieper R, Kleinkauf H and Zocher R (1992) Enniatin synthetases from different *Fusaria* exhibiting distinct amino acid specificities. Journal of Antibiotics 45: 1273–1277

Pieper R, Haese A, Schröder W and Zocher R (1995) Arrangement of catalytic sites in the multifunctional enzyme enniatin synthetase. European Journal of Biochemistry 230: 119–126

Shemyakin MM, Ovchinnikov YA, Ivanov NT, Antonov VK, Vinogradova EI, Shkrob AM, Malenkov GG, Evstratov AV, Laine IA, Melnik and Eland Ryabova ID (1969) Cyclodepsipeptides as chemical tools for studying transport through membranes. Journal of Membrane Biology 1: 402–430

Zocher R and Keller U (1997) Thiol template peptide synthesis systems in bacteria and fungi. Advances in Microbial Physiology 38: 85–131

Zocher R (2000) Purification of D-hydroxyisovalerate dehydrogenase form *Fusarium sambucinum*. Methods in Enzymology 324: 293–301

European Journal of Plant Pathology **108:** 719–734, 2002.

Secretion of natural and synthetic toxic compounds from filamentous fungi by membrane transporters of the ATP-binding cassette and major facilitator superfamily

Ioannis Stergiopoulos, Lute-Harm Zwiers and Maarten A. De Waard*
*Laboratory of Phytopathology, Department of Plant Sciences, Wageningen University, Binnenhaven 5, 6705 PD, P.O. Box 8025, 6700 EE Wageningen, The Netherlands; *Author for correspondence (Phone: +31317483123; Fax: +31317483412; E-mail: maarten.dewaard@fyto.dpw.wag-ur.nl)*

Key words: antibiotic compounds, fungicides, mycotoxin, pathogenicity, plant defence compounds, resistance, host-specific toxin

Abstract

This review provides an overview of members of the ATP-binding cassette (ABC) and major facilitator superfamily (MFS) of transporters identified in filamentous fungi. The most common function of these membrane proteins is to provide protection against natural toxic compounds present in the environment of fungi, such as antibiotics produced by other microorganisms. In plant pathogenic fungi, these transporters can also be an important determinant of virulence on host plants by providing protection against plant defence compounds or mediating the secretion of host-specific toxins. Furthermore, they play a critical role in determining base-line sensitivity to fungicides and other antimycotic agents. Overexpression of some of these transporters can lead to the development of resistance to chemically-unrelated compounds, a phenomenon described as multidrug resistance (MDR). This has been observed in a variety of organisms and can impose a serious threat to the effective control of pathogenic fungi.

Abbreviations: ABC – ATP-binding cassette; Acc. – accession number; DAS – diacetoxyscirpenol; DMI – demethylation inhibitor; EST – expressed sequence tag; MDR – multidrug resistance; MFS – major facilitator superfamily; MRP – multidrug resistance-related protein; NBF – nucleotide-binding fold; PDR – pleiotropic drug resistance; PDREs – pleiotropic drug resistance elements; SBI – sterol biosynthesis inhibitor; TMD – trans-membrane domain.

Introduction

In nature, filamentous fungi are constantly exposed to a wide variety of toxic compounds originating from various sources. In their living environment they encounter numerous antibiotic compounds produced by other microorganisms. Plant pathogenic fungi must also be able to resist the toxic effects of plant defence products such as phytoalexins and phytoanticipins. In addition, the advent of chemical disease control over the past decades led to an increased exposure of fungi to fungicides. On the other hand, fungi may also need to handle toxicants of endogenous origin such as antibiotics and mycotoxins that provide the producing organism with a competitive advantage in its ecological habitat. In this context, the following questions can be raised: 'Which strategies have organisms developed to protect themselves against exposure to exogenous toxic products?' and 'How do toxin producing organisms manage to protect themselves against auto-toxicity of endogenous toxins?' The answers to both questions relate to mechanisms of selective toxicity and can be of a qualitative or quantitative nature. Qualitative factors relate to the presence of the target site of a toxic compound in sensitive organisms and its absence in resistant ones. Quantitative factors determine the effective concentration of the cytotoxic agent that can be built up at the target site as a result of uptake, transport, storage and natural metabolism. Differences in affinity of the target site in target and non-target organisms are also important quantitative factors.

Functions of ABC and MFS transporters in filamentous fungi

Quantitative factors that influence the accumulation of toxicants in cells were only discovered in the last couple of decades and are described as members of the ATP-binding cassette (ABC) and the major facilitator superfamily (MFS) of membrane transporters. These transporters have a remarkably broad substrate specificity and are able to transport a wide variety of natural and synthetic toxic products of either endogenous or exogenous origin. In this way, they prevent or reduce accumulation of these products inside cells and hence, avoid or minimise their toxic action (Nelissen et al., 1997; Bauer et al., 1999; Pao et al., 1998). Furthermore, overexpression of these transporters is known to play an essential role in resistance of cells to chemically-unrelated compounds. This phenomenon is described as multidrug resistance (MDR) and is a serious threat to the effectiveness of drugs (Fling et al., 1991; Juliano and Ling, 1976).

The physiological functions of ABC and MFS transporters in fungi are also highly significant. These transporters provide protection against toxic compounds present in their natural habitat or prevent the cytoplasmatic accumulation of toxic secondary metabolites produced by the fungus itself. In addition, in plant pathogenic fungi these transporters may act as virulence or pathogenicity factors if they mediate secretion of virulence factors, such as host-specific toxins, or provide protection against plant defence compounds during pathogenesis (De Waard, 1997) (Figure 1). If the sexual stage of a fungus is present, ABC transporters can also be involved in secretion of mating type factors and thus, contribute to the genetic diversity of a fungal population. Such transporters have already been characterised in *Saccharomyces cerevisiae* and *Schizosaccharomyces pombe* (Christensen et al., 1997; Kuchler et al., 1989).

ABC and MFS transporters also function in protection against synthetic toxic compounds, such as fungicides and other antimycotic agents (De Waard, 1997). Over the past few years, the widespread use of antifungal compounds has resulted in the development of fungicide resistance in several fungal species. Studies on azole-resistant strains of *Candida albicans* and other fungal pathogens of medical importance indicated a decreased accumulation of these compounds in cells, mediated by ABC and MFS transporters. In many of these cases, exposure to a single antimycotic agent resulted in simultaneous resistance to a number

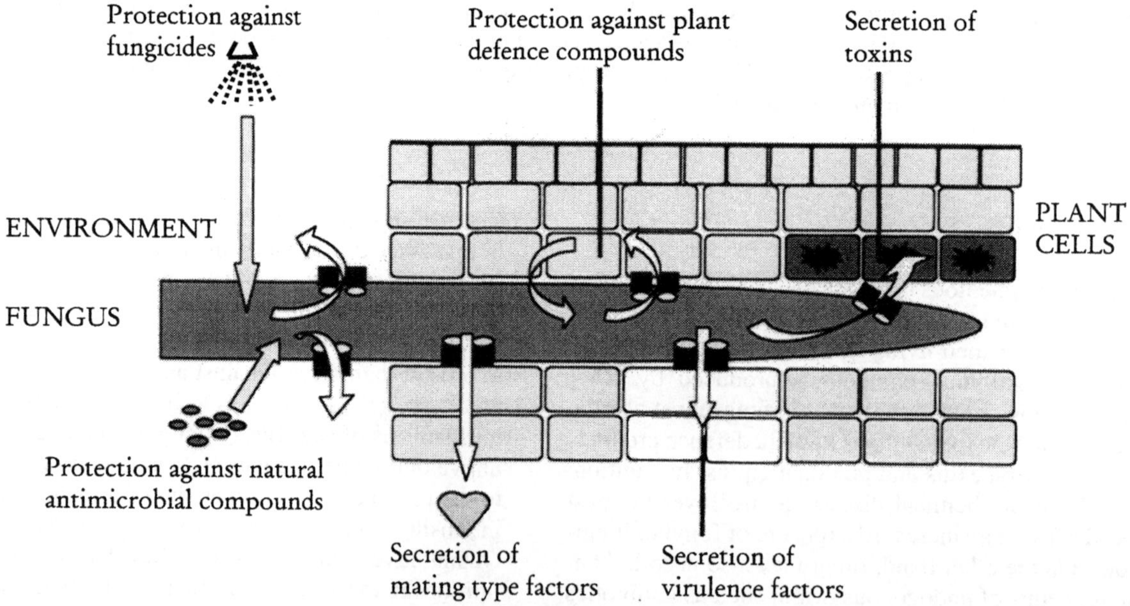

Figure 1. Functions of ABC and MFS transporters in filamentous fungi.

of chemically unrelated compounds. Overexpression of ABC and MFS genes was closely linked with such MDR phenotypes (Albertson et al., 1996; Fling et al., 1991; Franz et al., 1998; Sanglard et al., 1995; 1997; White, 1997). Reduced accumulation of azole fungicides in *Aspergillus nidulans* and *Penicillium italicum* suggested that a similar mechanism attributed to resistance development in these fungi (De Waard and Van Nistelrooy, 1980; 1984a,b; 1988). Recent studies support the hypothesis that ABC and MFS transporters are indeed involved in MDR in filamentous fungi (Del Sorbo et al., 2000).

The aim of this review is to describe ABC and MFS transporters from filamentous fungi, with a function in protection against natural toxic compounds and fungicides. This review will also focus on recent data that demonstrate the role of ABC and MFS transporters in secretion of host-specific toxins and mycotoxins.

ABC transporters

ABC transporters comprise one of the largest protein families known to date, operating in a wide variety of organisms from bacteria to man (Higgins, 1992). They are located in the outer plasma membrane or in membranes of intracellular compartments, such as the vacuoles, endoplasmatic reticulum, peroxisomes, and mitochondria. ABC transporters are capable of transporting a wide variety of cytotoxic, hydrophobic agents, ranging from ions to macromolecules, against a concentration gradient (Ambudkar et al., 1999; Bauer et al., 1999; Del Sorbo et al., 2000; Theodoulou, 2000). The energy needed for transport is generated by the hydrolysis of ATP and for this reason ABC transporters are characterised as primary active transport systems (Azzaria et al., 1989). ABC transporters are also described as P-glycoprotein, PDR (pleiotropic drug resistance) proteins or MDR proteins.

The structural unit of an ABC transporter is composed of two homologous halves each containing six trans-membrane domains (TMDs) and a conserved nucleotide-binding fold (NBF). The NBFs of ABC transporters are located in the cytoplasm. They are distinguished by the presence of highly conserved amino acid sequences, called the Walker A [G-(X)4-G-K-(T)-(X)6-I/V] and Walker B [R/K-(X)3-G-(X)3-L-(hydrophobic)4-D] motif, and the ABC signature [L-S-G-G-(X)3-R-hydrophobic-X-hydrophobic-A] (Ames et al., 1989; Walker et al., 1982). The catalytic activity of these sites with respect to coupling and hydrolysis of ATP provides the energy necessary for transport of substrates (Azzaria et al., 1989). Both NBFs can have ATPase activity, stimulated by the presence of the substrate and in the case of full size transporters interaction between the two homologous halves seems to be necessary for transport (al Shawi and Senior, 1993; Ambudkar et al., 1992; Loo and Clarke, 1994). The TMDs of ABC proteins are less conserved as compared to the NBFs. They might form a pore across the lipid bilayer of membranes (Rosenberg et al., 1997) and are known to play a role in determining the substrate specificity of the transporters. More specific, TMDs 4, 5, 6, 10, 11, and 12 and the extra-cellular loops connecting them are thought to be closely linked with substrate binding and transport (Greenberger, 1993; Loo and Clarke, 1995; Safa et al., 1990; Zhang et al., 1995).

ABC transporters can be classified into different clusters based on their topology. The majority of these proteins have a $[TMD_6\text{-}NBF]_2$ or $[NBF\text{-}TMD_6]_2$ topology. Half-sized transporters with a single TMD_6-NBF or NBF-TMD_6 configuration have also been described and are assumed to function after dimerisation (Decottignies and Goffeau, 1997; Theodoulou, 2000). Multidrug resistance-related proteins (MRP) are ABC transporters with a $TMD_n[TMD_6\text{-}NBF]_2$ topology. They are characterised by the presence of an additional trans-membrane spanning domain of approximately 200 amino acids at the N-terminus of the protein and the presence of a putative 'Regulatory' (R) or 'Connector' domain between the two homologue halves, thought to act in regulation of the protein (Tusnady et al., 1997) (Figure 2). Some representatives of this group of transporters have been identified as glutathione S-conjugate pumps involved in cellular detoxification and other processes (Ishikawa, 1992; Ishikawa et al., 1997).

ABC transporters include both uptake and efflux systems. In general, they exhibit a broad substrate specificity, although transporters with specific substrates also occur. The broad range of substrates for these proteins includes alkaloids, lipids, peptides, steroids, sterols, terpenoids, flavanoids, sugars, inorganic anions and heavy metal chelates. Synthetic compounds, such as fungicides, anticancer drugs, and other therapeutic or disease control agents have also been described (Ambudkar et al., 1999; Bauer et al., 1999; Del Sorbo et al., 2000; Theodoulou, 2000). Most of these compounds have a positive charge at physiological pH, are hydrophobic and enter the cells through passive diffusion (Gottesman and Pastan, 1993). The way that these compounds are transported is not

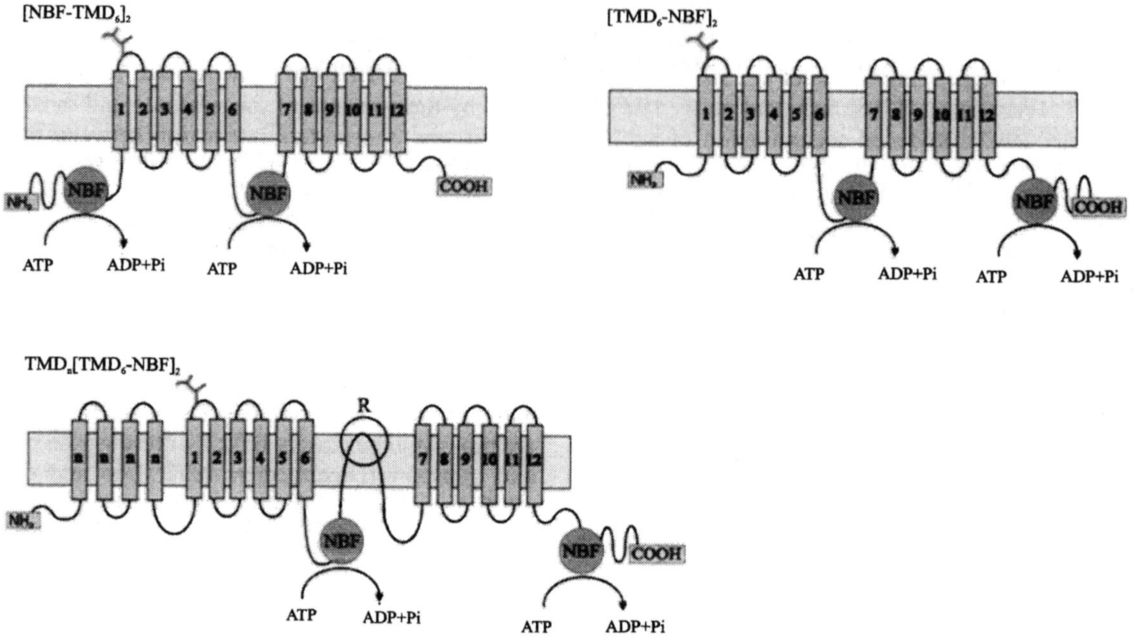

Figure 2. Schematic representation of ABC transporters with the $[NBF\text{-}TMD_6]_2$ and $[TMD_6\text{-}NBF]_2$ topology. MRP proteins are ABC transporters with a $TMD_n[TMD_6\text{-}NBF]_2$ topology. TMDs are indicated as boxes and are numbered. The NBFs are indicated as circles. 'R' is the 'regulatory' or 'connector' domain of the MRP-like proteins.

fully understood and several models have been proposed. Early models suggested that ABC proteins act as 'biological pumps' interacting directly with their substrates for their removal from the cytoplasm (Gottesman and Pastan, 1988). However, recent studies indicate that detection and excretion of toxic agents takes place at the membrane level before toxic concentrations can be built up in the cytoplasm. In this case, drugs are removed directly from the membranes into the extra-cellular space, and therefore, ABC transporters have been described as 'hydrophobic membrane vacuum cleaners' (Gottesman and Pastan, 1993; Raviv et al., 1990). However, it is also assumed that, initially, a small amount of the drug might be necessary to reach the cytoplasm in order to activate the cell defence mechanism and induce production of these pumps. A second model suggests that ABC transporters have a flippase activity, translocating drugs from the inner leaflet of the lipid membrane bilayer to the outer one and subsequently released into the outer environment (Higgins and Gottesman, 1992). Yet, other models propose that ABC proteins indirectly promote decreased intracellular accumulation of drugs by altering biophysical properties of membranes (Roepe, 1994; Wadkins, 1997).

In silico analysis of the *S. cerevisiae* genome identified at least 30 different ABC transporters (Taglicht and Michaelis, 1998). Other genome sequencing programmes reported the presence of 56 ABC proteins in *Drosophila melanogaster*, 56 in *Caenorhabditis elegans*, 129 in *Arabidopsis thaliana,* and 48 in *Homo sapiens* (http://www.humanabc.org/). Evidence suggests that an equally high number of ABC proteins are present in filamentous fungi. Complete sequencing of the *Neurospora crassa* genome revealed the existence of 39 putative ABC proteins, while in *Cochliobolus heterostrophus, Fusarium sporotrichioides*, and *Botrytis cinerea* 51, 54, and 46 ABC proteins were identified, respectively (Yoder and Turgeon, 2001). The function of a limited number of these proteins has been elucidated, but the vast majority remains to be investigated.

ABC transporters from filamentous fungi involved in protection against natural toxic compounds and fungicides

Aspergillus spp.

Several ABC transporter genes from *A. nidulans* have been described (Table 1). *AtrA* and *atrB* encode

Table 1. ABC transporter genes from *A. nidulans*

ABC gene	Acc.	Reference
atrA	Z68904	Del Sorbo et al., 1997
atrB	Z68905	Del Sorbo et al., 1997
atrC	AF071410	Andrade et al., 2000b
*atrC2**	AF082072	Angermayr et al., 1999
atrD	AF071410	Andrade et al., 2000b
atrE	AJ309280	Andrade et al., 2000
atrF	AJ309281	Andrade et al., 2000
atrG	AJ309282	Andrade et al., 2000
abcA	—	Do Nascimento et al., 1999
abcB	—	Do Nascimento et al., 1999
abcC	—	Do Nascimento et al., 1999
abcD	—	Do Nascimento et al., 1999

*Renamed. Original name was also *atrC*.

proteins with a [NBF-TMD_6]$_2$ topology, while *atrC* and *atrD* encode proteins with a [TMD_6-NBF]$_2$ topology. Expression of these genes is up-regulated by a range of natural and synthetic toxic compounds, such as the secondary plant metabolites pisatin and reserpine, the antibiotic cycloheximide, and sterol-demethylation-inhibiting (DMI) fungicides (Andrade et al., 2000; Del Sorbo et al., 1997). Heterologous expression of *atrB* in an ABC transporter deficient mutant strain of *S. cerevisiae* showed that yeast transformants carrying this gene are resistant to the antibiotic cycloheximide and a number of other drugs, suggesting a role for atrB in MDR (Del Sorbo et al., 1997). Functional analysis by gene replacement in *A. nidulans* demonstrated that atrB is involved in protection against compounds from all major classes of fungicides and natural toxic compounds. These include anilinopyrimide, benzimidazole, phenylpyrrole, phenylpyridylamine, DMI, and strobilurin fungicides, as well as the plant alkaloid camptothecin and the phytoalexin resveratrol (Andrade et al., 2000a). Disruption of *atrD* in *A. nidulans* resulted in a hypersensitive phenotype to cycloheximide, the cyclosporin derivative PSC 833, nigericin, and valinomycin (Andrade et al., 2000b). The results show that some of the *A. nidulans* ABC transporters play a role in defence against a wide range of natural toxic products and fungicides.

Since ABC transporters can play a role in protection of fungi against the activity of DMI fungicides, efforts have been undertaken to identify ABC transporter genes in *A. flavus* and *A. fumigatus*, two important opportunistic pathogens involved in human aspergillosis. *AflMDR1* (Acc. U62931) is the only ABC transporter gene cloned from *A. flavus*. It encodes a transporter with a [TMD_6-NBF]$_2$ predicted topology but no function for this protein has been reported yet. ABC transporter genes cloned from *A. fumigatus* are listed in Table 2. *AfuMDR1* encodes a protein with a predicted [TMD_6-NBF]$_2$ topology while *abcA* and *atrF* encode proteins with a [NBF-TMD_6]$_2$ topology. Interestingly, *AfuMDR2* encodes a protein with only four putative TMDs and a single NBF. Heterologous expression of AfuMDR1 in *S. cerevisiae* resulted in enhanced resistance to cilofungin, a (1,3)-β-D glucan synthase inhibitor, but not to other antimycotics tested. *A. fumigatus* is insensitive to cilofungin, suggesting that this property is due to the ability of AfuMDR1 to provide protection against this compound (Tobin et al., 1997). Increased levels of *ADR1* expression were found in an itraconazole-resistant isolate of *A. fumigatus* that does not accumulate this compound, implying the involvement of ADR1 in itraconazole efflux (Slaven et al., 1999). Reduced cellular accumulation of this compound has also been described for other itraconazole-resistant isolates of this fungus (Denning et al., 1997). Hence, increased drug efflux activity seems to be a common mechanism of resistance in *A. fumigatus*.

Table 2. ABC transporter genes from *A. fumigatus*

ABC gene	Acc.	Reference
AfuMDR1	U62934	Tobin et al., 1997
AfuMDR2	U62936	Tobin et al., 1997
ADR1	—	Slaven et al., 1999
abcA	AJ417501	—
atrF	AJ311940	—

Botrytis cinerea

B. cinerea (teleomorph *Botryotinia fuckeliana*) is the causal agent of the grey mould disease that attacks a wide variety of crop plants and causes serious economic losses. Several ABC transporter genes have been cloned from this fungus (Table 3). The basal level of transcripts of these genes vary from undetectable (*BcatrC*, *BcatrJ*, *BcatrN*), to low (*BcatrA*, *BcAtrB*, *BcatrE*, *BcatrG*, *BcatrK*), and high (*BcatrF*, *BcatrH*, *BcatrI*) (Vermeulen et al., 2001). Treatment with fungicides can increase transcript levels of several of these genes. *BcatrB* encodes a protein with a [NBF-TMD_6]$_2$ topology. Increased transcript levels of this gene are observed after treatment with phytoalexins, phenylpyrrole, anilinopyrimide, and dicarboximide fungicides (Schoonbeek et al., 2001; Vermeulen et al., 2001). Functional analysis by means of gene

Table 3. ABC transporter genes from *B. cinerea*

ABC gene	Acc.	Reference
BcatrA	Z68906	Del Sorbo and De Waard, 1996
BcatrB	AJ006217	Schoonbeek et al., 1999
BcatrC	AF241315	Vermeulen et al., 2001
BcatrD	AJ272521	Vermeulen et al., 2001
BcatrE	AF238224	Vermeulen et al., 2001
BcatrF	AF238230	Vermeulen et al., 2001
BcatrG	AJ278038	Vermeulen et al., 2001
BcatrH	AF241313	Vermeulen et al., 2001
BcatrI	AF238229	Vermeulen et al., 2001
BcatrJ	AF238228	Vermeulen et al., 2001
*BcatrK**	AF238227	Vermeulen et al., 2001
*BMR1**	AB028872	Nakajima et al., 2001
BcatrL	—	Vermeulen et al., 2001
BcatrM	—	Vermeulen et al., 2001
BcatrN	AF238226	Vermeulen et al., 2001

**BcatrK* and *BMR1* are identical.

disruption showed that Δ*BcatrB* mutants display increased sensitivity to resveratrol, a plant defence product in grapevine, while virulence tests showed a slight reduction in virulence on grapevine leaves as compared to the wild-type control. These results suggest a role for BcatrB in virulence of *B. cinerea* by providing protection against resveratrol. In addition, Δ*BcatrB* mutants exhibit increased sensitivity to the phenylpyrrole fungicides fenpiconil and fludioxonil, while mutants overexpressing this gene show decreased sensitivity to these compounds, suggesting an additional role for BcatrB in fungicide sensitivity of *B. cinerea* (Schoonbeek et al., 2001). *BcatrD* encodes a protein with a $[NBF\text{-}TMD_6]_2$ topology. This gene exhibits a high level of basal expression in germlings of *B. cinerea* and its transcript level is up-regulated by treatment with DMI, dicarboximide, and benzimidazole fungicides, as well as with the antibiotic cycloheximide (Hayashi et al., 2001). A positive correlation between increased transcript levels of *BcatrD* and resistance to azole fungicides was observed. Replacement mutants of *BcatrD* exhibit increased sensitivity to several DMIs and accumulate relatively high amounts of oxpoconazole. Likewise, mutants overexpressing *BcatrD* show a positive correlation between *BcatrD* expression level and decreased sensitivity to this compound. These results indicate that *BcatrD* is a determinant of sensitivity of *B. cinerea* to DMI fungicides (Hayashi et al., 2001). *BMR1* (*BcatrK*) encodes a protein with a $[NBF\text{-}TMD_6]_2$ topology. Δ*BMR1* mutants display increased sensitivity to the antibiotic polyoxin and the organophosphorous fungicide iprobenfos, which implies that BMR1 is an additional MDR transporter of this fungus (Nakajima et al., 2001).

Gibberella pulicaris

The necrotrophic fungus *G. pulicaris* (anamorph *Fusarium sambucinum*), the cause of dry rot disease of potatoes, infects tubers through open wounds. Recently, the ABC transporter gene *Gpabc1* (Acc. AJ306607), encoding a protein with a $[NBF\text{-}TMD_6]_2$ topology, was cloned and characterised from this fungus. Wounding of potato tissues induces the production of phytoalexins, such as rishitin and lubimin. Treatment of the fungus with either of these two compounds induced rapid expression of *Gpabc1*. Pathogenicity tests revealed that Δ*Gpabc1* mutants are unable to colonise potato slices. Additionally, these mutants are hypersensitive to rishitin and lubimin, suggesting that they are incapable of coping with the toxic effect of phytoalexins produced at the infection site. Thus, Gpabc1 is essential for pathogenicity of *G. pulicaris* on potato tubers by providing protection against plant defence compounds (Fleissner et al., 2002).

Magnaporthe grisea

M. grisea is a major pathogen of rice. The ABC transporter gene *ABC1* (Acc. AF032443) was identified through an insertional mutagenesis screen for pathogenicity mutants. The encoded protein has a $[NBF\text{-}TMD_6]_2$ topology. Δ*ABC1* mutants have normal growth on agar media but showed a complete loss of virulence on barley and rice plants. Histopathological analysis of the infection process on rice showed that the Δ*ABC1* mutants, although capable of forming appressoria, failed to produce extensive infection hyphae and died shortly after penetration of the epidermal cells. Expression analysis of the Δ*ABC1* mutants after treatment with several compounds demonstrated that *ABC1* transcript levels are strongly induced by the DMI fungicides miconazole and metconazole, the rice phytoalexin sakuranetin, and the protein synthesis inhibitor hygromycin. Yet, Δ*ABC1* mutants do not show increased sensitivity to any of these compounds. Thus, the exact role of ABC1 during pathogenesis needs to be established. The most probable explanation for the loss of virulence of the deletion mutants is that ABC1 provides protection against antimicrobial compounds present in barley and rice cells, although the compound(s) involved remain to be identified (Urban et al., 1999).

Mycosphaerella graminicola

Five ABC transporter genes have been cloned and sequenced from *M. graminicola* (anamorph *Septoria tritici*) the causal agent of *Septoria tritici* leaf blotch, one of the most important diseases of wheat (Table 4). The encoded ABC proteins all exhibit the $[NBF\text{-}TMD_6]_2$ configuration. Expression of *MgAtr3* was not detected under any conditions tested. However, *MgAtr1*, *MgAtr2*, *MgAtr4*, and *MgAtr5* display distinct expression profiles when treated with a range of compounds known to be either substrates or inducers of ABC transporters. These include DMIs, natural toxic compounds such as the plant defence compounds eugenol and psoralen, and the antibiotics cycloheximide and neomycin. The expression pattern of the genes also depends on the morphological state, yeast-like cells or mycelium, of the fungus (Stergiopoulos et al., 2002a; Zwiers and De Waard, 2000). Heterologous expression of *MgAtr1*, *MgAtr2*, *MgAtr4*, and *MgAtr5* genes in a multiple knockout strain of *S. cerevisiae* showed that the products of these genes transport a wide range of chemically-unrelated compounds and possess an extensive overlap in substrate specificity. Their substrate range includes synthetic compounds such as DMIs, natural compounds such as the plant metabolites berberine and camptothecin, and the mycotoxin diacetoxyscirpenol (DAS). The function of *MgAtr1-5* in virulence of *M. graminicola* on wheat was investigated with knockout mutants. Analysis of the transformants showed that Δ*MgAtr5* mutants have a small increase in sensitivity to the putative wheat defence compound resorcinol, suggesting a role for this transporter during pathogenesis. No further phenotypes were observed for any of the mutants and compounds tested. This could be due to redundancy of ABC transporters with similar substrate specificity (Zwiers, 2002). Thus, the possibility that some of these transporters are involved in protection against natural and synthetic toxic compounds cannot be excluded. All transformants were tested for virulence on wheat seedlings. Δ*MgAtr4* mutants displayed reduced virulence as compared to the wild-type control strain. Northern analysis on interaction RNA isolated from wheat infected with the wild-type isolate and the Δ*MgAtr4* mutant showed a low build-up of biomass of the Δ*MgAtr4* mutant on wheat (Stergiopoulos et al., 2002b). The findings indicate a role for this protein in virulence.

Table 4. ABC transporter genes from *M. graminicola*

ABC gene	Acc.	Reference
MgAtr1	AJ243112	Zwiers and De Waard, 2000
MgAtr2	AJ243113	Zwiers and De Waard, 2000
MgAtr3	AF364105	Stergiopoulos et al., 2002a
MgAtr4	AF329852	Stergiopoulos et al., 2002a
MgAtr5	AF364104	Stergiopoulos et al., 2002a

Penicillium digitatum

P. digitatum is the causal agent of the citrus green mould. Five ABC transporter genes have been cloned from this fungus (Table 5). *PMR1* was cloned from a DMI-resistant isolate that is also cross-resistant to the unrelated chemicals cycloheximide, 4-nitroquinoline-*N*-oxide (4NQO), and acriflavine. The gene has a $[NBF\text{-}TMD_6]_2$ topology and shares a high degree of identity with *atrA* and *atrB* from *A. nidulans*. Basal level of *PMR1* transcription was higher in DMI-resistant isolates of *P. digitatum* than in sensitive ones. Treatment of the fungus with the DMI fungicide triflumizole induced *PMR1* transcription in sensitive and DMI-resistant strains. Disruption of *PMR1* in sensitive and DMI-resistant isolates resulted in increased sensitivity to DMIs as well as to camptothecin, phloretin, and oligomycin (Nakaune et al., 1998; 2002). Reintroduction of *PMR1* in Δ*PMR1* mutants derived from DMI-resistant strains resulted in decreased sensitivity to DMIs. However, reintroduction of *PMR1* in Δ*PMR1* mutants derived from DMI-sensitive strains did not result in decreased sensitivity. These results suggest that although PMR1-mediated efflux of DMIs plays an important role in determining sensitivity of *P. digitatum* to DMIs, it does not, alone, explain the difference in sensitivity between DMI-sensitive and resistant isolates (Hamamoto et al., 2000). *PMR5* encodes a protein with a $[NBF\text{-}TMD_6]_2$ topology and shares a high degree of identity with *PMR1* from *P. digitatum*, *atrB* from *A. nidulans* and *BcatrB* from *B. cinerea*. In contrast to *PMR1*, transcription of *PMR5* is strongly induced by benzimidazoles, dithianon, and resveratrol but not by DMIs. Furthermore, Δ*PMR5* mutants display increased sensitivity to benzimidazole fungicides and dithianon as well as to plant products such as

Table 5. ABC transporter genes from *P. digitatum*

ABC gene	Acc.	Reference
PMR1	AB010442	Nakaune et al., 1998
PMR3	—	Nakaune et al., 2001
PMR4	—	Nakaune et al., 2001
PMR5	AB060639	Nakaune et al., 2001; 2002

camptothecin and resveratrol. The results demonstrate that PMR5 and PMR1 possess distinct substrate specificity and have an important role in providing protection of the fungus against a range of natural and synthetic toxic compounds (Nakaune et al., 2002).

Other filamentous fungi

ABC transporter genes from other fungi include two genes, coded *LMABC1* and *LMABC2,* from *Leptosphaeria maculans* (anamorph *Phoma lingam*), the causal agent of blackleg disease of crucifers. Both genes encode proteins with the $[NBF\text{-}TMD_6]_2$ topology. *LMABC1* is strongly induced after treatment of the fungus with an analogue of the brassinin phytoalexin methyl-4-chlorobenzyldithiocarbamate and the azole fungicide miconazole. Increased transcripts of *LMABC2* are observed after treatment of the fungus with cycloheximide and the phytotoxin sirodesmin PL. Functional complementation of a *S. cerevisiae* ABC transporter null mutant, demonstrated that LMABC2 is able of transporting cycloheximide and 4NQO. Thus, it is possible that this gene plays a role in MDR, thereby protecting the fungus against natural and synthetic toxic compounds (Taylor and Condie, 1999).

Four ABC transporter genes have been cloned from *Venturia inaequalis*, the causal agent of the apple scab disease (Table 6). *ViABC1* and *ViABC2* encode proteins with a $[NBF\text{-}TMD_6]_2$ topology, while *ViABC3* and *ViABC4* encode proteins with a $[TMD_6\text{-}NBF]_2$ topology. Northern blot analysis revealed that *ViABC4* has a high basal level of expression indicating that this gene might be involved in basic metabolism (Schnabel and Jones, 2001).

ABC transporters involved in transport of toxins

Although data on the role of fungal ABC proteins in the secretion of endogenous produced toxins are limited, it is clear that ABC transporters can at least provide protection against toxins produced by other fungi. Bissinger and Kuchler (1994) reported the first evidence that ABC transporters can function in protection against mycotoxins. They showed that *PDR5* (Acc. L19922), a well-known ABC transporter gene from *S. cerevisiae*, can provide protection against sporidesmin, an epidithidioxopiperazine mycotoxin produced by the fungus *Phytomyces chartarum*. Δ*PDR5* mutants are supersensitive to sporidesmin as well as to a number of unrelated drugs, suggesting that PDR5 is involved in cellular detoxification processes. Trichothecene mycotoxins can act as virulence factors (Desjardins et al., 1996). The expression level of *PDR5* in *S. cerevisiae* correlates with sensitivity to exogenous trichothecenes. Transgenic tobacco plants transformed with *PDR5* were less sensitive to the trichothecene 4,15-diacetoxyscirpenol (Muhitch et al., 2000).

Table 6. ABC transporter genes from *V. inaequalis*

ABC gene	Acc.	Reference
ViABC1	AF227914	Schnabel et al., 2001
ViABC2	AF227915	Schnabel et al., 2001
ViABC3	AF375878	Schnabel et al., 2001
ViABC4	AF375879	Schnabel et al., 2001

Preliminary data have been reported for sirodesmins, which are non-selective toxins produced by the plant pathogenic fungus *L. maculans*. Increased expression levels of *LMABC2* are found after treatment with sirodesmin. The fungus is also known to produce the host-specific phytotoxin phomalide, suggesting that LMABC2 is involved in secretion of such compounds (Taylor and Condie, 1999). Further functional analysis of these ABC genes with respect to secretion of toxins is under investigation.

Heterologous expression of *MgAtr1* and *MgAtr4* from *M. graminicola* in a *S. cerevisiae* strain with multiple non-functional ABC genes, demonstrated that both genes confer decreased sensitivity to the trichothecene DAS, a mycotoxin produced by *Fusarium graminearum*. Kema et al. (1996) proposed that formation of necrotic lesions on wheat leaves infected with *M. graminicola* is associated with secretion of phytotoxic compounds produced by the pathogen. Although such compounds have not yet been characterised, it might be possible that MgAtr1 and MgAtr4 are involved in secretion of host-specific toxins produced by the fungus. The reduced virulence phenotype of the Δ*MgAtr4* mutants supports this hypothesis (Stergiopoulos et al., 2002b).

Furthermore, it has been shown that MRP transporters of *H. sapiens* are capable of energy-dependent transport of aflatoxin B1 and its glutathione conjugates (Loe et al., 1997). These reports suggest that protection of organisms against aflatoxins can be mediated via efflux activity of ABC transporters.

Structure–function relationships among ABC transporters

We performed a phylogenetic analysis of ABC transporters from filamentous fungi sharing the

[NBF-TMD$_6$]$_2$ topology. Multiple sequence alignments were made using the ClustalW programme and a dendrogram was generated by parsimony as calculated using PROTPARS (Figure 3). A cluster of ABC proteins involved in protection against natural toxic compounds can be distinguished. This cluster includes atrB from *A. nidulans,* BcatrB from *B. cinerea*, MgAtr5 from *M. graminicola*, and PMR5 from *P. digitatum.* These transporters are induced and/or transport plant-defence related compounds, with the stilbene resveratrol and the plant alkaloid camptothecin as the most striking examples. It is possible that this branch of ABC transporters represents a group of orthologs with a similar function. A second cluster seems to be constituted by ABC proteins involved in protection against DMIs. This cluster includes BcatrD from *B. cinerea*, atrE and atrG from *A. nidulans*, and PMR1 from *P. digitatum.* BcatrD and PMR1 are specifically involved in protection against synthetic toxic compounds such as DMIs, while *atrG* and *atrE* are induced by DMIs (Andrade, 2000). Similarities in smaller clusters can also be found. For example, Gpabc1 from *G. pulicaris* and ABC1 from *M. grisea* are both pathogenicity factors. MgAtr1 from *M. graminicola* and BcatrA from *B. cinerea* cluster together and are both induced by cycloheximide. The same holds true for atrA from *A. nidulans* and MgAtr4 from *M. graminicola* that are both strongly induced by cycloheximide and imazalil.

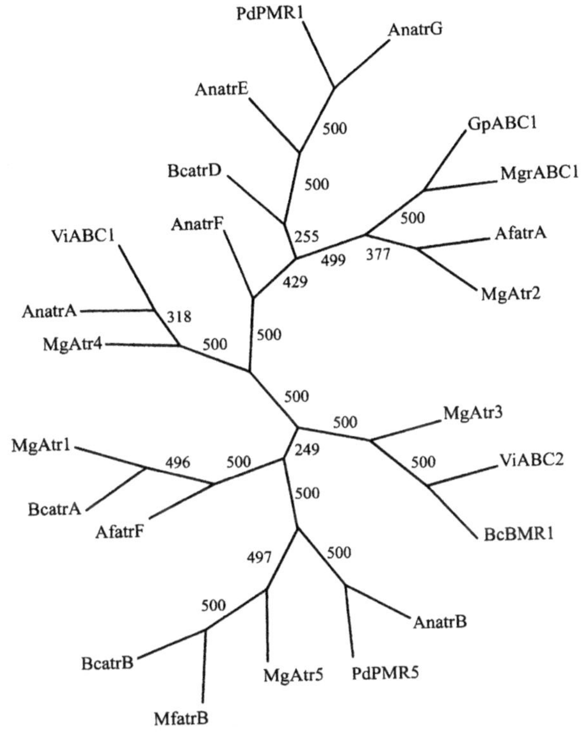

Figure 3. Phylogenetic unrooted tree of 23 ABC transporters with the [NBF-TMD$_6$]$_2$ topology from filamentous fungi. The tree was based on parsimony as calculated using PROTPARS. Alignments were performed using the ClustalW alignment programme. The code of the proteins mentioned in the tree are proceeded by two or three letters, indicating the source organism: An: *Aspergillus nidulans*, Af: *Aspergillus fumigatus*, Bc: *Botrytis cinerea*, Gp: *Gibberella pulicaris*, Mf: *Monilinia fructicola*, Mg: *Mycosphaerella graminicola*, Mgr: *Magnaporthe grisea*, Pd: *Penicillium digitatum*, Vi: *Venturia inaequalis*.

MFS transporters

MFS transporters comprise the largest protein family, present from bacteria to higher eukaryotes. They facilitate the uniport, symport, or antiport of various compounds using the energy from electrochemical gradients across membranes. For this reason MFS transporters are classified as secondary transport systems (Marger and Saier, 1993; Pao et al., 1998; Paulsen et al., 1996).

In general, MFS transporters consist of 12 or 14 TMDs arranged into two homologous halves, joined together by a large cytoplasmatic loop between TMDs 6 and 7 (Henderson, 1993; Kilty and Amara, 1992; Paulsen and Skurray, 1993) (Figure 4). Similarities between the two halves of the protein suggest that MFS transporters with 12 TMDs evolved

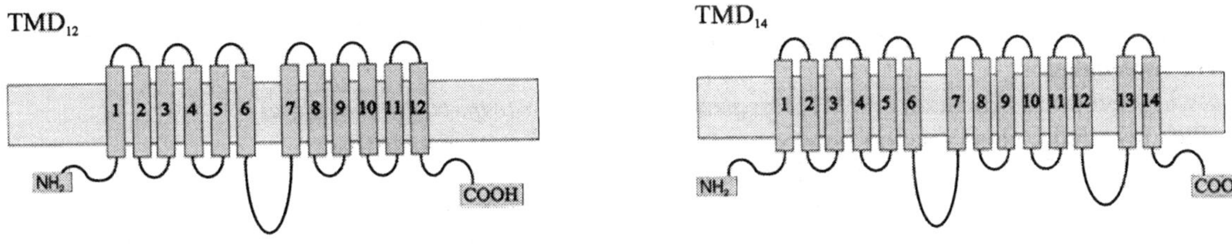

Figure 4. Schematic representation of MFS transporters with 12 and 14 TMDs. The TMDs are indicated in boxes and are numbered.

from a duplication event of a gene encoding a protein with 6 TMDs, while transporters with 14 TMDs emerged by additional acquisition of 2 TMDs at the C-terminal domain of the protein (Pao et al., 1998; Paulsen and Skurray, 1993; Rubin et al., 1990). Mutational analysis with respect to functionality of the two halves of these proteins suggests that the N-terminal domain is primarily involved in proton translocation while the C-terminal domain is engaged in substrate binding and recognition (Griffith et al., 1992). Unlike ABC proteins, MFS transporters do not possess well-defined conserved motifs. An overview of this superfamily classified at least 18 distinct families. Sequence alignment identified the presence of a conserved motif of 13 amino acids between TMDs 2 and 3 [G-[RKPATY]-L-[GAS]-[DN]-[RK]-[FY]-G-R-[RK]-[RKP]-[LIVGST]-[LIM]]. This motif is thought to be involved in promoting conformational changes in the protein upon substrate binding, allowing trafficking of substrates through the membrane. The motif may also act as a gate, regulating substrate transport from and to the cytoplasm. Additional sequence motives present in members of specific families of the MFS transporters can also be found, although the significance and conservation of such motives among the different MFS families is not yet fully defined (Pao et al., 1998).

Although functional analysis of most MFS transporters is still in its initial stages, reports suggest that the function and substrate specificity of these proteins can be impressively broad. MFS transporters are involved in the trafficking of sugars, drugs, polyols, vitamins, neurotransmitters, Krebs-cycle metabolites, phosphorylated glycolytic intermediates, amino acids, peptides, osmolites, iron-siderophores, nucleosides and organic and inorganic anions and cations. In this way, they facilitate cell functions such as the uptake of nutrients, secretion of cell cycle metabolites, protection against endogenous and exogenous toxic compounds, and maintenance of an electrochemical gradient across membranes. They are also involved in sporulation of fungi and yeasts, cell to cell communication and pathogenesis. MFS transporters differ from ABC proteins by the fact that they only mediate transport of relatively small-sized molecules.

The phylogenetic classification of MFS proteins into families revealed a close relation between structure and function. Thus, members of different families have specific substrate specificity and predictions about the function of novel members of a specific family can be made (Nelissen et al., 1997; Pao et al., 1998; Paulsen et al., 1996). In this respect, transporters of sugars and drugs comprise by far the largest families. For example, genome analysis of *S. cerevisiae* revealed 186 MFS transporters. A total of 28 of these proteins play a role in MDR while 34 were classified as sugar transporters (Nelissen et al., 1997).

MFS transporters involved in secretion of host-specific toxins

The first MFS transporter involved in secretion of host-specific toxins was reported by Pitkin et al. (1996). They found that the *TOXA* (Acc. AAB36607) gene product from *Cochliobolus carbonum*, encoding a MFS transporter with 10–13 predicted TMDs, secretes the host-specific cyclic tetrapeptide HC-toxin. Interestingly, *TOXA* occurs in two linked copies flanking the *HTS1* gene that encodes the central enzyme in HC-toxin biosynthesis. This suggests that there is a cluster of genes responsible for HC-toxin production. The *TOXA* genes have only been found in fungal strains that produce the HC-toxin. Mutants with a single-disrupted copy of *TOXA* still produce the HC-toxin and are virulent on maize. However, attempts to disrupt both copies of the *TOXA* gene were unsuccessful, suggesting that the encoded protein acts in self-protection against the HC-toxin. Thus, TOXA is essential for survival and virulence on host plants of HC-toxin producing strains.

Many species of the fungal genus *Cercospora* spp. including the soybean pathogen *C. kikuchii*, produce the phytotoxic polyketide cercosporin. Sequencing of cDNA clones generated from mRNA transcripts that were specifically induced under light-conditions led to the identification of *CFP* (Acc. AAC78076). This gene encodes a MFS transporter with 14 predicted TMDs that is involved in cercosporin transport. Δ*CFP* mutants do not produce cercosporin, have reduced virulence on soybean, and show increased sensitivity to this toxin. Complementation of these mutants with a functional *CFP* copy restores the parental phenotype. The results indicate that CFP is a cercosporin transporter involved in virulence and self-protection against this toxin (Callahan et al., 1999; Upchurch et al., 2001).

MFS transporters involved in secretion of mycotoxins

Most of the genes involved in trichothecene biosynthesis by *Fusarium* spp. are located within a gene cluster. In *F. sporotrichioides*, this cluster includes

Tri12 (Acc. AF11355), an MFS gene encoding a transporter with 14 predicted TMDs. Mutants of *S. cerevisiae* transformed with *Tri12* have decreased sensitivity to trichothecene. Disruption of *Tri12* in *F. sporotrichioides* reduced secretion of trichothecene and the Δ*Tri12* mutants showed impaired *in vitro* growth. Thus, Tri12 acts as a trichothecene transporter and plays a role in self-protection of *F. sporotrichioides* against trichothecenes (Alexander et al., 1999).

Recently, the *A. flavus* MFS gene *aflT* was identified in a gene cluster involved in aflatoxin biosynthesis (Acc. AC087725). The protein encoded by *aflT* is highly homologous to the HC-toxin transporter TOXA from *C. carbonum*. These observations suggest that *aflT* encodes an aflatoxin transporter (Chang et al., 1999).

The information available, indicates that secretion of endogenously-produced mycotoxins proceeds via MFS transporters, while protection against exogenous trichothecenes and aflatoxins seems to be mediated via ABC transporters. This is a striking difference, which might relate to the evolution of these proteins.

MFS transporters with other functions

Fragments of three MFS genes from *B. cinerea* were cloned from an expressed sequence tag (EST) library of this fungus grown under nitrogen starvation conditions. These genes were coded *Bcmfs1* (Acc. AF238225), *Bcmfs2* (Acc. AF241312), and *Bcmfs4* (Acc. AF238231). *Bcmfs1* encodes a protein with 14 predicted TMDs, which has highest homology to the MFS transporters Apafl1 from *A. parasiticus*, CFP1 from *C. kikuchii*, and TOXA from *C. carbonum*. DMI-resistant strains of *B. cinerea* show an increased expression of *Bcmfs1*. Δ*Bcmfs1* mutants display an increased sensitivity to the plant defence compound camptothecin and the toxin cercosporin. Mutants overexpressing *Bcmfs1* show an increased tolerance to these natural toxic compounds as well as to several DMIs, the dicarboximide fungicide iprodione, and the fungicides fenhexamide and captan. Deletion of *Bcmfs1* in a Δ*BcatrD* mutant increased the sensitivity of this mutant to DMIs. These results demonstrate that *Bcmfs1* is a multidrug transporter. Bcmfs1 is the first fungal MFS transporter identified with a function in MDR, capable of transporting both natural toxic compounds and fungicides (Hayashi et al., 2002).

In many cases, fungal MFS transporters involved in secretion of host-specific toxins are located in gene clusters responsible for the biosynthesis of these products. However, this is not always true as the *Gibberella fujikuroi* MFS transporter gene *smt* (Acc. AJ272424) is located adjacent to the gibberellin (GA) biosynthesis pathway genes but not involved in secretion of GAs from this fungus. *Smt* is probably a member of the sugar-transporter family as expression of this gene is induced by sugar alcohols, such as sorbitol, mannitol, and myoinositol (Voss et al., 2001).

Concluding remarks

State of the art

In eukaryotes, all well-characterised drug efflux mechanisms can be ascribed to the activity of ABC and MFS transporters. These transporters belong to superfamilies that are large and ancient, being thought to date back more that 3 billion years. It is suggested that the subfamilies of drug transporters evolved from families with other transport capacities and that members of these parent families may have functioned as transporters of nutrients into the cell or as exporters of biosynthetic macromolecules (Saier et al., 1998).

The ABC transporters characterised in filamentous fungi seem to be specifically involved in transport of exogenous substrates. These can be toxic compounds of either natural or synthetic origin. The rapid production of these transporters upon exposure to cytotoxic agents suggests that these proteins function as a first line of defence for the survival of the fungus. This protection mechanism is highly significant, as ABC transporters possess a broad range of substrate specificity and thus, can provide protection against toxic compounds of different chemical classes. This is a fundamental difference compared to other defence mechanisms, which are effective only against a particular class of compounds or even a single molecule. In plant pathogenic fungi, ABC transporters also provide protection against plant defence compounds, and thus, act as virulence factors. No ABC transporters that secrete endogenously produced virulence factors such as host-specific toxins have yet been identified. ABC transporters from filamentous fungi also have an important role in influencing base-line sensitivity to fungicides. Furthermore, overproduction of ABC proteins can result in MDR to different classes of fungicides. This has been reported especially for pathogens of medical importance (St Georgiev, 2000; Vanden Bossche et al., 1998). Yet, the risk of MDR in fungal pathogens of agricultural importance seems to be limited (De Waard et al., 1995). This discrepancy is ascribed to a relatively low fitness of MDR strains by which they are unable to compete with wild-type isolates under field conditions (De Waard et al., 1982; Dekker, 1981).

Transporters of endogenous toxins most frequently belong to the MFS transporters. This may be explained by the fact that transport of such substrates should be constitutive as endogenously synthesised toxicants can be produced during the major part of the lifetime of the fungus. In this case, maintenance of MFS transport systems may be more cost-effective than the ABC transporter systems. MFS proteins are also involved in transport of exogenous substrates, such as cytotoxic compounds and thus, can function as a protection mechanism of cells. These MFS transporters are usually members of a specific 12 TMDs subfamily and are also implicated in MDR. Hence, activation of different efflux mechanisms seems necessary to ensure successful protection of cells from toxicants. For example, in a recent study on fluconazole-resistant strains of *C. albicans*, overexpression of ABC and MFS transporter genes operated concurrently in 85% of all the resistant strains tested (Perea et al., 2001).

Practical implications

The observation that ABC and MFS transporters influence base-line sensitivity to fungicides, are responsible for MDR and act as virulence factors implies that these transporters constitute an attractive target for chemical control. In this context, inhibitors of ABC and MFS transporters may improve the efficacy of chemical control and reduce virulence of plant pathogenic fungi.

Inhibition of ABC transporter activity can occur through compounds that are termed chemosensitizers or MDR modulators. These modulators may have little or no intrinsic cytotoxic action but inhibit ABC transporter-mediated drug export through competitive inhibition of transport or through interaction with other binding sites of the transporter. Compounds that block the generation of ATP in cells can also inhibit the activity of the ABC proteins. This approach has already been validated in clinical trials aimed at the reversal of MDR in tumor cells (Avendano and Menendez, 2002). Mixtures of agricultural fungicides and ABC transporter modulators may also be synergistic. This has been demonstrated for mixtures of DMIs and respiratory inhibitors, such as oligomycin and dicyclohexylcarbodiimide (De Waard and Van Nistelrooy, 1982; 1984a,b).

This review has described how ABC proteins with high sequence homologies can have similar functions. This observation implies that it should be feasible to develop inhibitors of specific ABC transporters. Such targeted inhibition is of particular interest for the cluster of DMI transporters and for the ones involved in virulence of plant pathogens. Inhibition of specific ABC transporters could also result in selectivity between target and non-target organisms, which is, for instance, required during plant pathogenesis.

Efflux mechanisms can interfere with screening programmes aimed at the discovery of new fungitoxic lead compounds. The reason is that test organisms have an intrinsic insensitivity to test compounds as a consequence of efflux systems that reduce accumulation of the compounds in cells. In this case, promising lead compounds would be missed. Thus, the use of hypersensitive fungal strains in high-throughput screening processes for the identification of new antifungal compounds is important. Single and multiple knockout mutants of ABC genes with a hypersensitive phenotype could be very useful for this purpose. Furthermore, to prevent fungicide resistance development based upon increased efflux activity, candidate compounds should not act as substrates of ABC transporters. This implies that these compounds must have the same toxic activity to wild-type, disruption, and overexpression mutants of ABC transporters. Hence, such mutants should also be included in the optimisation of new fungicides.

Many fungal species are known to produce antibiotics and other important metabolic compounds. Secretion of such compounds by the producing organisms might be mediated via efflux pumps such as ABC and MFS transporters. For example, penicillin secretion by *A. nidulans* may be influenced by the ABC transporter atrD (Andrade et al., 2000b). This implies that the production of secondary metabolites could be increased by overexpression of specific transporters.

Future perspectives

The number of ABC and MFS transporters cloned from filamentous fungi rapidly increases and genome-sequencing programmes will provide an even larger amount of data regarding sequences of ABC and MFS transporters. Although the role of some of these proteins has been elucidated, the function of the majority of transporters is still based on speculations derived from sequence homologies and expression data. Additional data on phenotypic characterisation of gene disruption and/or overexpression mutants are certainly needed.

In many cases, ABC and MFS proteins have a remarkable broad substrate specificity that may overlap. This abundance of transporters presents serious difficulties in the functional analysis of gene disruption mutants. For this reason, overexpression mutants might be more helpful in the characterisation of the substrate specificity of the transporters.

Expression data show that ABC or MFS transporter genes can respond to the same stimuli, such as cytotoxic agents. This might suggest a general stress response and/or a common regulatory mechanism for these transporters. Yet, the way in which fungi regulate expression of ABC or MFS transporters is still unknown. Several studies have indicated that the promoter activity of the human P-glycoprotein may be modulated via the so-called Ras signal transduction pathway (Bosch and Croop, 1996). No evidence is available for the presence in filamentous fungi of a signal transduction pathway or a specific transcription factor that could regulate transporter activity. The way cells perceive potential substrates may be closely related with such a transduction pathway. Specific substrate receptors in membranes or membrane disturbance caused by the presence of the substrate might possibly lead to activation of such a transduction pathway. Alternatively, it might be possible that the transporter itself is involved in recognition of substrates and thus regulate its own expression.

Regulation of ABC transporter gene expression in *S. cerevisiae* is described as the PDR network. In this system, expression of several ABC transporters is under the regulatory control of PDR1 and PDR3. These proteins are members of the bi-nuclear Gal4p-like $Zn(II)_2Cys_6$ class of transcription factors that regulate transcription through *cis*-acting elements. These elements (5′-TCCG/aC/tGG/cA/g-3′) were coded as PDREs (pleiotropic drug resistance elements) and are present in all known PDR1/PDR3 target sites including MFS and ABC transporter genes (Bauer et al., 1999; Wolfger et al., 2001). Thus, it would be interesting to search for PDR1/PDR3 homologues in filamentous fungi and for putative PDREs elements. In *M. graminicola* disruption of single ABC genes did not result in increased sensitivity to any of the compounds tested. However, heterologous expression studies in *S. cerevisiae* shows that the encoded proteins posses an overlapping specificity in transport of substrates and northern blot analysis reveals that they are also induced by several common compounds (Stergiopoulos et al., 2002a; Zwiers, 2002; Zwiers and De Waard, 2000). Such observations support the idea that regulation of these transporters might proceed via a common transduction pathway.

An interesting question remains how ABC and MFS transporters or putative cell membrane receptors recognise and transport compounds from diverse chemical classes. This is especially relevant for xenobiotics since these were only developed in the last three decades. Despite this, several ABC and MFS transporters effectively transport these compounds. This may be due to a low substrate specificity of multidrug transporters. However, it is also possible that synthetic compounds mimic natural analogues. For example, DMI fungicides structurally resemble sterols and for this reason might be perceived as such by cells.

In summary, fungal ABC and MFS transporters present an exciting field of research since it combines fundamental aspects with various practical applications. Advances in molecular biology techniques especially with respect to fungal transformation and high-throughput functional analysis will strongly stimulate this type of research for the near future and may lead to the discovery of new disease control agents.

Acknowledgements

We kindly acknowledge Prof. Dr. P.J.G.M. de Wit for critical reading of the manuscript. I. Stergiopoulos is financially supported by the Training and Mobility of Researchers (TMR) Programme – Marie Curie Research Training Grants, The European Commission (Contract No. ERBFMBICT983558) and L.-H. Zwiers by the Graduate School of Experimental Plant Sciences (EPS), Wageningen University, the Netherlands.

References

al Shawi MK and Senior AE (1993) Characterization of the adenosine triphosphatase activity of Chinese hamster P-glycoprotein. Journal of Biological Chemistry 268: 4197–4206

Albertson G, Niimi M, Cannon R and Jenkinson H (1996) Multiple efflux mechanisms are involved in *Candida albicans* fluconazole resistance. Antimicrobial Agents and Chemotherapy 40: 2835–2841

Alexander NJ, McCormick SP and Hohn TM (1999) TRI12, a trichothecene efflux pump from *Fusarium sporotrichioides*: gene isolation and expression in yeast. Molecular and General Genetics 261: 977–984

Ambudkar S, Lelong I, Zhang J, Cardarelli C, Gottesman M and Pastan I (1992) Partial Purification and Reconstitution of the Human Multidrug-Resistance Pump: Characterization of the Drug-Stimulatable ATP Hydrolysis. Proceedings of the National Academy of Sciences of United States of America 89: 8472–8476

Ambudkar SV, Dey S, Hrycyna CA, Ramachandra M, Pastan I and Gottesman MM (1999) Biochemical, cellular, and pharmacological aspects of the multidrug transporter. Annual Review of Pharmacology and Toxicology: 39361–39398

Ames GFL, Nikaido K, Groarke J and Petithory J (1989) Reconstitution of periplasmic transport in inside-out membrane vesicles: energization by ATP. Journal of Biological Chemistry 264: 3998–4002

Andrade AC (2000) ABC transporters and multidrug resistance in *Aspergillus nidulans*. PhD Thesis Wageningen University, Wageningen, the Netherlands, ISBN 90-5808-279-2

Andrade AC, Del Sorbo G, Van Nistelrooy JGM and De Waard MA (2000a) The ABC transporter AtrB from *Aspergillus nidulans* mediates resistance to all major classes of fungicides and some natural toxic compounds. Microbiology 146: 1987–1997

Andrade AC, Van Nistelrooy JGM, Peery RB, Skatrud PL and De Waard MA (2000b) The role of ABC transporters from *Aspergillus nidulans* in protection against cytotoxic agents and in antibiotic production. Molecular and General Genetics 263: 966–977

Avendano C and Menendez JC (2002) Inhibitors of multidrug resistance to antitumor agents (MDR). Current Medical Chemistry 9: 159–193

Azzaria M, Schurr E and Gros P (1989) Discrete mutations introduced in the predicted nucleotide-binding sites of the *mdr1* gene abolish its ability to confer multidrug resistance. Molecular and Cellular Biology 9: 5289–5297

Bauer BE, Wolfger H and Kuchler K (1999) Inventory and function of yeast ABC proteins: about sex, stress, pleiotropic drug and heavy metal resistance. Biochimica Biophysica Acta, Biomembranes: 217–236

Bissinger PH and Kuchler K (1994) Molecular cloning and expression of the *Saccharomyces cerevisiae* Sts1 gene product – a yeast ABC transporter conferring mycotoxin resistance. Journal of Biological Chemistry 269: 4180–4186

Bosch I and Croop J (1996) P-glycoprotein multidrug resistance and cancer. Biochimimica Biophysica Acta 1288: F37–F54

Callahan TM, Rose MS, Meade MJ, Ehrenshaft M and Upchurch RG (1999) CFP, the putative cercosporin transporter of *Cercospora kikuchii*, is required for wild type cercosporin production, resistance, and virulence on soybean. Molecular Plant–Microbe Interactions 12: 901–910

Chang P-K, Yu J, Bhatnagar D and Cleveland TE (1999) The carboxy-terminal portion of the aflatoxin pathway regulatory protein AFLR of *Aspergillus parasiticus* activates GAL1::lacZ gene expression in *Saccharomyces cerevisiae*. Applied and Environmental Microbiology 65: 2508–2512

Christensen PU, Davey J and Nielsen O (1997) The *Schizosaccharomyces pombe* mam1 gene encodes an ABC transporter mediating secretion of M-factor. Molecular and General Genetics 255: 226–236

De Waard MA, Van Nistelrooy JGM, Langeveld CR, Van Kan JAL and Del Sorbo G (1995) Multidrug resistance in filamentous fungi. 11th International Rheinhardsbrunn Symposium: Modern Fungicides and Antifungal Compounds (pp 293–300) Friedrichroda, Germany

De Waard MA (1997) Significance of ABC transporters in fungicide sensitivity and resistance. Pesticide Science 51: 271–275

De Waard MA, Groeneweg H and Van Nistelrooy JGM (1982) Laboratory resistance to fungicides which inhibit ergosterol biosynthesis in *Penicillium italicum*. Netherlands Journal of Plant Pathology: 99–112

De Waard MA and Van Nistelrooy JGM (1982) Antagonistic and synergistic activities of various chemicals on the toxicity of fenarimol to *Aspergillus nidulans*. Pesticide Science: 279–286

De Waard MA and Van Nistelrooy JGM (1980) An energy-dependent efflux mechanism for fenarimol in a wild-type strain and fenarimol-resistant mutants of *Aspergillus nidulans*. Pesticide Biochemistry Psysiology 13: 255–266

De Waard MA and Van Nistelrooy JGM (1984a) Differential accumulation of fenarimol by a wild-type isolate and fenarimol-resistant isolates of *Penicillium italicum*. Netherlands Journal of Plant Pathology: 143–153

De Waard MA and Van Nistelrooy JGM (1984b) Effects of phthalimide fungicides on the accumulation of fenarimol by *Aspergillus nidulans*. Pesticide Science: 56–62

De Waard MA and Van Nistelrooy JGM (1988) Accumulation of SBI fungicides in wild-type and fenarimol-resistant isolates of *Penicillium italicum*. Pesticide Science 22: 371–382

Decottignies A and Goffeau A (1997) Complete inventory of the yeast ABC proteins. Nature Genetics 15: 137–145

Dekker J (1981) Impact of fungicide resistance of disease control toxicity. British Crop Protection Conference: pests and diseases (pp 857–863) Brighton, England

Del Sorbo G, Andrade AC, Van Nistelrooy JG, Van Kan JA, Balzi E and De Waard MA (1997) Multidrug resistance in *Aspergillus nidulans* involves novel ATP-binding cassette transporters. Molecular and General Genetics 254: 417–426

Del Sorbo G, Schoonbeek H and De Waard MA (2000) Fungal transporters involved in efflux of natural toxic compounds and fungicides. Fungal Genetics and Biology 30: 1–15

Denning DW, Venkateswarlu K, Oakley KL, Anderson MJ, Manning NJ, Stevens DA, Warnock DW and Kelly SL (1997) Itraconazole resistance in *Aspergillus fumigatus*. Antimicrobial Agents and Chemotherapy 41: 1364–1368

Desjardins AE, Proctor RH, Bai G, McCormick SP, Shaner G, Buechley G and Hohn TM (1996) Reduced virulence of trichothecene-nonproducing mutants of Gibberella zeae in wheat field tests. Molecular Plant–Microbe Interactions 9: 775–781

Do Nascimiento AM, Terenzi MF, Goldman MMH and Goldman GH (1999) A novel ATP-binding cassette transporter involved in multidrug resistance in the filamentous fungus *Aspergillus nidulans*. Fungal Genetics Newsletter 46: 44

Fleissner A, Sopalla C and Weltring K (2002) An ATP-binding cassette multidrug-resistance transporter is necessary for tolerance of *Gibberella pulicaris* to phytoalexins and virulence on potato tubers. Molecular Plant–Microbe Interactions 15: 102–108

Fling ME, Kopf J, Tamarkin A, Gorman JA, Smith HA and Koltin Y (1991) Analysis of a *Candida albicans* gene that encodes a novel mechanism for resistance to benomyl and methotrexate. Molecular and General Genetics 227: 318–329

Franz R, Kelly SL, Lamb DC, Kelly DE, Ruhnke M and Morschhauser J (1998) Multiple molecular mechanisms contribute to a stepwise development of fluconazole resistance in clinical *Candida albicans* strains. Antimicrobial Agents and Chemotherapy 42: 3065–3072

Gottesman MM and Pastan I (1988) The multidrug transporter, a double-edged sword. Journal of Biological Chemistry 263: 12163–12166

Gottesman MM and Pastan I (1993) Biochemistry of multidrug resistance mediated by the multidrug transporter. Annual Review of Biochemistry 62: 385–427

Greenberger LM (1993) Major photoaffinity drug labeling sites for iodoaryl azidoprazosin in P-glycoprotein are within, or immediately C-terminal to, transmembrane domains 6 and 12. Journal of Biological Chemistry 268: 11417–11425

Griffith JK, Baker ME, Rouch DA, Page MG, Skurray RA, Paulsen IT, Chater KF, Baldwin SA and Henderson PJ (1992) Membrane transport proteins: implications of sequence comparisons. Current Opinion in Cell Biology 4: 684–695

Hamamoto H, Hasegawa K, Nakaune R, Lee YJ, Makizumi Y, Akutsu K and Hibi T (2000) Tandem repeat of a transcriptional enhancer upstream of the sterol 14alpha-demethylase gene (CYP51) in *Penicillium digitatum*. Applied and Environmental Microbiology 66: 3421–3426

Hayashi K, Schoonbeek H and De Waard MA (2002) *Bcmfs1*, a novel ABC transporter from *Botrytis cinerea*, provides tolerance to the natural toxic compounds camptothecin and to fungicides (submitted)

Hayashi K, Schoonbeek H, Sugiura H and De Waard MA (2001) Multidrug resistance in *Botrytis cinerea* associated with decreased accumulation of the azole fungicide oxpoconazole and increased transcription of the ABC transporter gene *BcatrD*. Pesticide Biochemistry and Physiology 70: 168–179

Henderson P (1993) The 12-transmembrane helix transporters. Current Opinion in Cell Biology 5: 708–721

Higgins CF and Gottesman M (1992) Is the multidrug transporter a flippase? Trends in Biochemical Sciences 17: 18–21

Higgins CF (1992) ABC transporters: from microorganisms to man. Annual Review of Cell Biology 8: 67–113

Ishikawa T (1992) The ATP-dependent glutathione S-conjugate export-pump. Trends in Biochemical Sciences 17: 463–468

Ishikawa T, Li ZS, Lu YP and Rea PA (1997) The GS-X pump in plant, yeast, and animal cells: structure, function, and gene expression. Bioscience Reports 17: 189–207

Juliano RL and Ling V (1976) A surface glycoprotein modulating drug permeability in Chinese hamster ovary cell mutants. Biochimica Biophysica Acta 455: 152–162

Kema GHJ, Yu D, Rijkenberg FHJ, Shaw MW and Baayen RP (1996) Histology of the pathogenesis of *Mycosphaerella graminicola* in wheat. Phytopathology 86: 777–786

Kilty JE and Amara SG (1992) Families of twelve transmembrane domain transporters. Current Opinion in Biotechnology 3: 675–682

Kuchler K, Sterne RE and Thorner J (1989) *Saccharomyces cerevisiae* STE6 gene product: A novel pathway for protein export in eukaryotic cells. The EMBO Journal 8: 3973–3984

Loe DW, Stewart RK, Massey TE, Deeley RG and Cole SP (1997) ATP-dependent transport of aflatoxin B1 and its glutathione conjugates by the product of the multidrug resistance protein (MRP) gene. Molecular Pharmacology 51: 1034–1041

Loo TW and Clarke DM (1994) Reconstitution of drug-stimulated ATPase activity following co-expression of each half of human P-glycoprotein as separate polypeptides. Journal of Biological Chemistry 269: 7750–7755

Loo TW and Clarke DM (1995) Covalent modification of human P-glycoprotein mutants containing a single cysteine in either nucleotide-binding fold abolishes drug-stimulated ATPase activity. Journal of Biological Chemistry 270: 22957–22961

Marger MD and Saier MHJ (1993) A major superfamily of transmembrane facilitators that catalyse uniport, symport and antiport. Trends in Biochemical Sciences 18: 13–20

Muhitch MJ, McCormick SP, Alexander NJ and Hohn TM (2000) Transgenic expression of the TRI101 or PDR5 gene increases resistance of tobacco to the phytotoxic effects of the trichothecene 4,15-diacetoxyscirpenol. Plant Science 157: 201–207

Nakajima M, Suzuki J, Hosaka T, Hibi T and Akutsu K (2001) Functional analysis of an ATP-binding cassette transporter gene in *Botrytis cinerea* by gene disruption. Journal of General Plant Pathology 67: 212–214

Nakaune R, Adachi K, Nawata O, Tomiyama M, Akutsu K and Hibi T (1998) A novel ATP-binding cassette transporter involved in multidrug resistance in the phytopathogenic fungus *Penicillium digitatum*. Applied and Environmental Microbiology 64: 3983–3988

Nakaune R, Hamamoto H, Imada J, Akutsu K and Hibi T (2002) A novel ABC transporter gene, *PMR5*, is involved in multidrug resistance in the phytopathogenic fungus *Penicillium digitatum*. Molecular Genetics and Genomics: Published online: 22 February

Nelissen B, De Wachter R and Goffeau A (1997) Classification of all putative permeases and other membrane plurispanners of the major facilitator superfamily encoded by the complete genome of *Saccharomyces cerevisiae*. FEMS Microbiology Reviews 21: 113–134

Pao SS, Paulsen IT and Saier MH, Jr. (1998) Major facilitator superfamily. Microbiology and Molecular Biology Reviews 62: 1–34

Paulsen IT, Brown MH and Skurray RA (1996) Proton-dependent multidrug efflux systems. Microbiology Reviews 60: 575–608

Paulsen IT and Skurray RA (1993) Topology, structure and evolution of two families of proteins involved in antibiotic and antiseptic resistance in eukaryotes and prokaryotes – an analysis. Gene 132: 155

Perea S, Lopez-Ribot JL, Kirkpatrick WR, McAtee RK, Santillan RA, Martinez M, Calabrese D, Sanglard D and Patterson TF (2001) Prevalence of molecular mechanisms of resistance to azole antifungal agents in *Candida albicans* strains displaying high-level fluconazole resistance isolated from human immunodeficiency virus-infected patients. Antimicrobial Agents and Chemotherapy 45: 2676–2684

Pitkin JW, Panaccione DG and Walton JD (1996) A putative cyclic peptide efflux pump encoded by the TOXA gene of the plant-pathogenic fungus *Cochliobolus carbonum*. Microbiology 142: 1557–1565

Raviv Y, Pollard H, Bruggemann E, Pastan I and Gottesman M (1990) Photosensitized labeling of a functional multidrug transporter in living drug-resistant tumor cells. Journal of Biological Chemistry 265: 3975–3980

Roepe P (1994) Indirect mechanism of drug transport by P-glycoprotein. Trends in Pharmacological Sciences 15: 445–446

Rosenberg MF, Callaghan R, Ford RC and Higgins CF (1997) Structure of the multidrug resistance P-glycoprotein to 2.5 nm resolution determined by electron microscopy and image analysis. Journal of Biological Chemistry 272: 10685–10694

Rubin RA, Levy SB, Heinrikson RL and Kezdy FJ (1990) Gene duplication in the evolution of the two complementing domains of gram-negative bacterial tetracycline efflux proteins. Gene 87: 7–13

Safa AR, Stern RK, Choi K, Agresti M, Tamai I, Mehta ND and Roninson IB (1990) Molecular basis of preferential resistance to colchicine in multidrug-resistant human cells conferred by Gly-185 fwdarw Val-185 substitution in P-glycoprotein. Proceedings of the National Academy of Sciences of the United States of America 87: 7225–7229

Saier MH, Jr., Paulsen IT, Sliwinski MK, Pao SS, Skurray RA and Nikaido H (1998) Evolutionary origins of multidrug and drug-specific efflux pumps in bacteria. The FASEB Journal 12: 265–274

Sanglard D, Ischer F, Monod M and Bille J (1997) Cloning of *Candida albicans* genes conferring resistance to azole antifungal agents: Characterization of CDR2, a new multidrug ABC transporter gene. Microbiology 143: 405–416

Sanglard D, Kuchler K, Ischer F, Pagani JL, Monod M and Bille J (1995) Mechanisms of resistance to azole antifungal agents in *Candida albicans* isolates from AIDS patients involve specific multidrug transporters. Antimicrobial Agents and Chemotherapy 39: 2378–2386

Schnabel G and Jones LA (2001) Fungicide resistance genes in *Venturia inaequalis*. Modern fungicides and antifungal compounds. 13th International Reinhardsbrunn Symposium: Modern Fungicides and Antifungal Compounds, Friedrichroda, Germany

Schoonbeek H, Del Sorbo G and De Waard MA (2001) The ABC transporter BcatrB affects the sensitivity of *Botrytis cinerea* to the phytoalexin resveratrol and the fungicide fenpiclonil. Molecular Plant–Microbe Interactions 14: 562–571

Slaven JW, Anderson MJ, Sanglard D, Dixon GK, Bille J, Roberts IA and Denning DW (1999) Fungal Genetics Newsletter 46: 64

St Georgiev V (2000) Membrane transporters and antifungal drug resistance. Current Drug Targets 1: 261–284

Stergiopoulos I, Gielkens MMC, Goodall SD, Venema K and De Waard MA (2002a) ABC transporter genes from the wheat pathogen *Mycosphaerella graminicola*. Gene 289: 141–149

Stergiopoulos I, Zwiers L-H and De Waard MA (2002b) ABC transporters involved in pathogenesis of *Mycosphaerella graminicola*. 6th European Conference of Fungal Genetics (p 250) Pisa, Italy

Taglicht D and Michaelis S (1998) *Saccharomyces cerevisiae* ABC proteins and their relevance to human health and disease. Methods in Enzymology 292: 130–162

Taylor JL and Condie J (1999) Characterisation of ABC transporters from the fungal phytopathogen *Leptosphaeria maculans*. 9th International Congress on Molecular Plant–Microbe Interactions (p L71) Amsterdam, the Netherlands

Theodoulou FL (2000) Plant ABC transporters. Biochimica Biophysica Acta, Biomembranes 1465: 79–103

Tobin MB, Peery RB and Skatrud PL (1997) Genes encoding multiple drug resistance-like proteins in *Aspergillus fumigatus* and *Aspergillus flavus*. Gene 200: 11–23

Tusnady GE, Bakos E, Varadi A and Sarkadi B (1997) Membrane topology distinguishes a subfamily of the ATP-binding cassette (ABC) transporters. FEBS Letters 402: 1–3

Upchurch RG, Rose MS and Eweida M (2001) Over-expression of the cercosporin facilitator protein, CFP, in *Cercospora kikuchii* up-regulates production and secretion of cercosporin. FEMS Microbiology Letters 204: 89–93

Urban M, Bhargava T and Hamer JE (1999) An ATP-driven efflux pump is a novel pathogenicity factor in rice blast disease. The EMBO Journal 18: 512–521

Vanden Bossche H, Dromer F, Improvisi I, Lozano-Chiu M, Rex JH and Sanglard D (1998) Antifungal drug resistance in pathogenic fungi. Medical Mycology 36: 119–128

Vermeulen T, Schoonbeek H and De Waard MA (2001) The ABC transporter BcatrB from *Botrytis cinerea* is a determinant of the phenylpyrrole fungicide fludioxonil. Pest Management Science 57: 393–402

Voss T, Schulte J and Tudzynski B (2001) A new MFS-transporter gene next to the gibberellin biosynthesis gene cluster of *Gibberella fujikuroi* is not involved in gibberellin secretion. Current Genetics 39: 377–383

Wadkins RM RP (1997) Biophysical aspects of P-glycoprotein-mediated multidrug resistance. International Review of Cytology 171: 121–165

Walker JE, Saraste M, Runswick MJ and Gay NJ (1982) Distantly related sequences in the alpha- and beta-subunits of ATP synthase, myosin, kinases and other ATP-requiring enzymes and a common nucleotide binding fold. The EMBO Journal 1: 945–951

White T (1997) Increased mRNA levels of ERG16, CDR, and MDR1 correlate with increases in azole resistance in *Candida albicans* isolates from a patient infected with human immunodeficiency virus. Antimicrobial Agents and Chemotherapy 41: 1482–1487

Wolfger H, Mamnun YM and Kuchler K (2001) Fungal ABC proteins: pleiotropic drug resistance, stress response and cellular detoxification. Research in Microbiology 152: 375–389

Yoder OC and Turgeon BG (2001) Fungal genomics and pathogenicity. Current Opinion in Plant Biology 4: 315–321

Zhang XP, Collins KI and Greenberger LM (1995) Functional evidence that transmembrane 12 and the loop between transmembrane 11 and 12 form part of the drug-binding domain in P-glycoprotein encoded by MDR1. Journal of Biological Chemistry: 5441–5448

Zwiers L-H (2002) ABC transporters of the wheat pathogen *Mycosphaerella graminicola*. PhD Thesis Wageningen University, Wageningen, the Netherlands, ISBN 90-5808-613-5

Zwiers LH and De Waard MA (2000) Characterization of the ABC transporter genes *MgAtr1* and *MgAtr2* from the wheat pathogen *Mycosphaerella graminicola*. Fungal Genetics and Biology 30: 115–125